**DMV Seminar
Band 32**

# Ten Lectures on
# Random Media

**Erwin Bolthausen**
**Alain-Sol Sznitman**

Springer Basel AG

Authors:

Erwin Bolthausen
Institute for Applied Mathematics
University of Zurich
Winterthurerstrasse 190
8057 Zürich
Switzerland

e-mail: eb@amath.unizh.ch

Alain-Sol Sznitman
Department of Mathematics
ETH Zentrum
8092 Zürich
Switzerland

e-mail: sznitman@math.ethz.ch

2000 Mathematical Subject Classification 60G50, 60F05, 60J25, 82B41

A CIP catalogue record for this book is available from the
Library of Congress, Washington D.C., USA

Deutsche Bibliothek Cataloging-in-Publication Data
Bolthausen, Erwin:
Ten lectures on random media / Erwin Bolthausen ; Alain-Sol Sznitman. -
Basel ; Boston ; Berlin : Birkhäuser, 2002
   (DMV-Seminar ; Bd. 32)

ISBN 978-3-7643-6703-9        ISBN 978-3-0348-8159-3 (eBook)
DOI 10.1007/978-3-0348-8159-3

© 2002 Springer Basel AG

Originally published by Birkhäuser Verlag, Basel, Switzerland in 2002

Printed on acid-free paper produced from chlorine-free pulp. TCF ∞
Cover design: Heinz Hiltbrunner, Basel

9 8 7 6 5 4 3 2 1

# Ten Lectures on Random Media: A Short Presentation

The field of random media has been the object of an intensive mathematical activity over the last twenty-five years. It gathers a variety of models generally originating from physical sciences such as condensed matter physics, physical chemistry, biophysics, geology, and others, where certain materials have defects or inhomogeneities. This feature can be taken into account by letting the medium be random. This randomness in the medium turns out to cause very unexpected effects, especially in the large scale behavior of some of these models. What often in the beginning was deemed to be a simple toy-model, ended up as a major mathematical challenge. After now more than twenty years of intensive research in this field, certain new paradigms and some general methods have emerged, and the surprising results on the asymptotic behavior of individual models are now better understood in more general frameworks.

In these "Ten Lectures on Random Media" we try to give an account of some of these developments. However the present monograph by no means offers a complete overview of the field. For instance, we completely leave out percolation theory or random Schrödinger operators, to mention only two. The lectures concentrate on random motions in random media, and on mean-field spin glasses. In the case of random motions in random media the lectures in particular discuss one of the general methods, the so-called "point of view of the environment viewed from the particle", and one paradigm which has recently emerged, namely the "preponderant role of atypical pockets of low principal eigenvalues". In the case of mean-field spin glasses the lectures mainly discuss the (generalized) random energy model (GREM), and some recently developed aspects of it. The Parisi theory predicts that many models exhibit in the limit a GREM-like structure, in particular the celebrated Sherrington-Kirkpatrick model, but most of mean-field spin glass theory is still very far from a mathematically rigorous understanding.

The material we discuss grew out of the DMV-lectures on Random Media, in November 1999 at the Mathematical Research Institute in Oberwolfach. It is a pleasure to thank Prof. Matthias Kreck, its Director, for inviting us to deliver these lectures, and for this very pleasant and stimulating week.

Zürich, May 2001

# Contents

# Contents

# PART ONE

by Alain-Sol Sznitman, ETH Zürich

## LECTURES ON RANDOM MOTIONS IN RANDOM MEDIA

## Foreword

The following notes grew out of lectures held during the DMV-Seminar on Random Media in November 1999 at the Mathematics Research Institute of Oberwolfach, and in February-March 2000 at the Ecole Normale Supérieure in Paris.

In both places the atmosphere was very friendly and stimulating. The positive response of the audience was encouragement enough to write up these notes. I hope they will carry over the enjoyment of the live lectures. I whole heartedly wish to thank Profs. Matthias Kreck and Jean-François Le Gall who were responsible for these two very enjoyable visits, Laurent Miclo for his comments on an earlier version of these notes, and last but not least Erwin Bolthausen who was my accomplice during the DMV-Seminar.

# A Brief Introduction

The main theme of this series of lectures are "Random motions in random media". The subject gathers a variety of probabilistic models often originated from physical sciences such as solid state physics, physical chemistry, oceanography, biophysics ..., in which typically some diffusion mechanism takes place in an inhomogeneous medium. Randomness appears at two levels. It comes in the description of the motion of the particle diffusing in the medium, this is a rather traditional point of view for probability theory; but it also comes in the very description of the medium in which the diffusion takes place. The mathematical appeal of the subject stems from the fact that making the environment random has far reaching consequences, and simply stated models display unforeseen behaviors which represent mathematical challenges.

We shall now describe some of these models. As a starting point, we begin with a model of diffusion in a constant medium, namely we consider the nearest neighbor random walk $(X_n)$ on $\mathbb{Z}$,

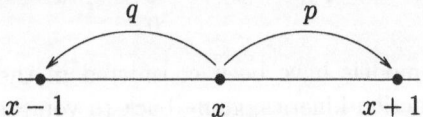

with fixed probabilities $p$ to jump to the right and $q$ to jump to the left. There are many different ways to introduce randomness in the medium and influence the nature of the diffusion taking place. Here are some examples.

## a) Site randomness:

One chooses i.i.d. variables $p(x, \omega)$, $x \in \mathbb{Z}$, with values in $[0, 1]$, and for a given realization of the environment, $(X_n)$ is now a Markov chain with probability $p(x, \omega)$ of jumping to the right neighbor $x + 1$, and $q(x, \omega) = 1 - p(x, \omega)$ of jumping to the left neighbor, given it is located in $x$ at time $n$:

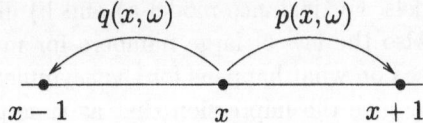

This model is the "random walk in random environment". It goes back to Chernov [10] and Temkin [70], and was originally introduced in the context of biochemistry to mimick the replication of DNA chains.

## b) Bond randomness:

One now chooses i.i.d. variables $c_{x,x+1}(\omega)$, $x \in \mathbb{Z}$, with values in $(0, \infty)$, and for a given realization of the environment, $(X_n)$ is a Markov chain with transition kernel determined as in a) by $p(x,\omega)$ and $q(x,\omega) = 1 - p(x,\omega)$, however now $p(x,\omega) = \frac{c_{x,x+1}(\omega)}{c_{x-1,x}(\omega)+c_{x,x+1}(\omega)}$.

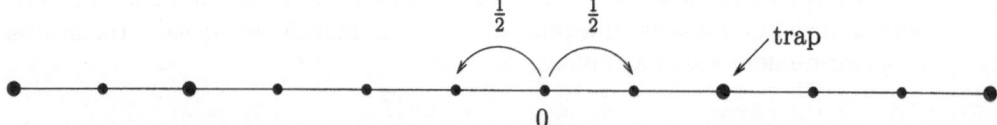

This is in essence a random conductivity model of the type introduced in disordered media physics. Such models can already be found in Fatt [22], Kirkpatrick [34], more references are provided in Chapter 5 of the book of Hughes [27].

**c) Random traps:**

One chooses i.i.d. Bernoulli variables $\alpha(x,\omega)$, $x \in \mathbb{Z}$, so that $\alpha(x,\omega) = 1$ signals the presence of a trap at $x$ and $\alpha(x,\omega) = 0$ the absence of a trap. The simple random walk is launched at the origin, and is "killed" when first meeting a trap.

Such type of trapping models have been considered in the context of physical chemistry to analyze reaction kinetics, going back to von Smoluchowski [72], but also in solid state physics, Rosenstock [55], [56]. Many references can for instance be found in the review article by Den Hollander-Weiss [26]. In fact some of the issues concerning this model are closely linked to the analysis of the Schrödinger equation with random potentials.

All the above models can easily be generalized to higher dimension. It is also possible to provide continuous models with similar features. However in this set of lectures we shall restrict the discussion to discrete models to somehow alleviate the level of technicality.

As mentioned above, unforeseen effects are unraveled by the investigation of these simply stated models. For instance model a) and b) display quite different asymptotic behaviors. Also the law of large numbers for model a) is quite surprising. A first guess based on what happens for the asymmetric nearest neighbor simple random walk may give the impression that as time goes on, the particle scans large portions of the medium so that a law of large numbers (at the level of the medium) should take over, and

$$\frac{X_n}{n} \longrightarrow \mathbb{E}[p] - \mathbb{E}[q], \text{ as } n \to \infty,$$

($\mathbb{P}$ denotes the probability governing the medium and $\mathbb{E}$ the corresponding expectation). As it turns out, the argument although containing some truth, (we shall see this in the next lecture) turns out to be wrong.

The law of large numbers for model a), when $d = 1$, goes back to the thesis [61] of Frank Spitzer's student Fred Solomon, who showed that depending on the relative location of 1 relative to the numbers $\frac{1}{\mathbb{E}[\rho]} < \mathbb{E}[\frac{1}{\rho}]$, with $\rho(x, \omega) = \frac{q(x,\omega)}{p(x,\omega)}$, (as follows from Jensen's inequality, when the one-site distribution is nondegenerate):

$$\text{if } 1 < \frac{1}{\mathbb{E}[\rho]}, \qquad \frac{X_n}{n} \longrightarrow v_\infty = \frac{1 - \mathbb{E}[\rho]}{1 + \mathbb{E}[\rho]} > 0, \qquad .$$

$$(\text{and } v_\infty < \mathbb{E}[p] - \mathbb{E}[q]),$$

$$(0.1) \qquad \text{if } \frac{1}{\mathbb{E}[\rho]} \leq 1 \leq \mathbb{E}\left[\frac{1}{\rho}\right], \quad \frac{X_n}{n} \longrightarrow v_\infty = 0,$$

$$\text{if } \mathbb{E}\left[\frac{1}{\rho}\right] < 1, \qquad \frac{X_n}{n} \longrightarrow v_\infty = \frac{\mathbb{E}[\frac{1}{\rho}] - 1}{\mathbb{E}[\frac{1}{\rho}] + 1} < 0,$$

$$(\text{and } v_\infty > \mathbb{E}[p] - \mathbb{E}[q]).$$

Thus the law of large numbers in this one-dimensional model is already quite intricate. The result also displays some tendency of the model to slowdown the walk. This is further exemplified by Sinai [60], who has shown that when $0 < \mathbb{E}[(\log \rho)^2] < \infty$, and $\mathbb{E}[\log \rho] = 0$, $X_n$ typically has size of order $(\log n)^2$ under $P_0 = \mathbb{P} \times P_{0,\omega}$ ($P_{0,\omega}$ is the law of the chain $(X_n)$ in the environment $\omega$, when starting from 0). In fact much more is known, cf. Sinai [60], Kesten [31], or the book by Revesz [54].

Although the subject of random motions in random media embodies many different models, we have tried to advertise two general ideas in the subsequent lectures. First we discuss mostly in Lecture 1 and 2, some elements of the "point of view of the environment viewed from the particle". This technique has turned out to be very helpful in a number of situations, in particular in the analysis of the model b) above. Then in Lecture 3 and 5 we advertise a paradigm present in several problems of random media, namely the "preponderant role of pockets of atypically low principal eigenvalues". In these problems, certain eigenvalues, which turn out to govern the asymptotics, exhibit substantial fluctuations, due to the very randomness of the medium, and some atypical relatively small pockets where the relevant eigenvalue is especially low, play a key role. Lecture 4 begins the discussion of random walks in random environment. Finally, the reader will quickly realize that many aspects of random motions in random media are not covered here. The following notes by no means represent an overview of the field, but rather reflect the current interests of the author.

# Lecture 1:

# The Environment Viewed from the Particle

In this lecture we shall briefly discuss some elements of the point of view of "the environment viewed from the particle". This set of ideas can be found in the work of S.M. Kozlov as well as Papanicolaou-Varadhan (cf. [36] or [48]), further developments are discussed in Kipnis-Varadhan [33], De Masi et al. [42], Olla [46] and the references therein. We begin with some notations.

The environment is described by the set $\Omega$, where:

in case of model a) of the introduction:

$\Omega = \mathcal{P}_\kappa^{\mathbf{Z}^d}$, where for $\kappa > 0$ some fixed number $\mathcal{P}_\kappa$ denotes the set of $(2d)$-vectors $(p(e))_{|e|=1, e \in \mathbf{Z}^d}$ with components in $[\kappa, 1]$ such that $\sum_{|e|=1} p(e) = 1$,

in case of model b) of the introduction:

$\Omega = [a, b]^{\mathcal{E}_d}$, with $\mathcal{E}_d = \{\{x, y\}; \, x, y \in \mathbf{Z}^d, |x - y| = 1\}$ the set of nearest neighbor bonds in $\mathbf{Z}^d$, and $0 < a < b$, are some fixed numbers.

$\Omega$ is tacitly endowed with the canonical product $\sigma$-field $\mathcal{B}$.

$\mathbb{P}$ is a product measure on $\Omega$, under which the canonical coordinates are i.i.d. variables.

$t_x$, $x \in \mathbf{Z}^d$, are the canonical translations on $\Omega$, which preserve $\mathbb{P}$.

$P_{x,\omega}$, for $x \in \mathbf{Z}^d$, $\omega \in \Omega$, denotes the law of the Markov chain in $\mathbf{Z}^d$, with transition probability:

$$p(y, e, \omega) = \begin{array}{l} \omega(y, e), \text{ in case a) (random walk in random environment)}, \\ \dfrac{\omega(\{y, y+e\})}{\sum_{|e'|=1} \omega(\{y, y+e'\})} \text{ in case b) (random conductivity model)}, \end{array}$$

for $y \in \mathbf{Z}^d$, $|e| = 1$.

Observe that $(\Omega, \mathcal{B}, (t_x)_{x \in \mathbf{Z}^d}, \mathbb{P})$ is ergodic and $p(x, e, \omega) = p(0, e, t_x \omega)$. The **environment viewed from the particle** is the $\Omega$-valued process:

(1.1) $$\overline{\omega}_n \stackrel{\text{def}}{=} t_{X_n} \omega, \, n \geq 0.$$

The first observation is:

**Proposition 1.1.** *Under $P_{0,\omega}$, $\omega \in \Omega$, (resp. $P_0 = \mathbb{P} \times P_{0,\omega}$), $(\overline{\omega}_n)$ is a Markov chain with state space $\Omega$, transition kernel:*

(1.2)     $Rf = \sum_{|e|=1} p(0, e, \omega)\, f \circ t_e$, *f bounded measurable on $\Omega$, and initial law $\delta_\omega$ (resp. $\mathbb{P}$).*

*Proof.* If $f_i, i = 0, \ldots, n+1$, are bounded measurable functions on $\Omega$,

(1.3)     $E_{0,\omega}[f_{n+1}(\overline{\omega}_{n+1})\, f_n(\overline{\omega}_n) \ldots f_0(\overline{\omega}_0)] \overset{(1.1)}{=} E_{0,\omega}[f_{n+1}(t_{X_{n+1}}\omega) \ldots f_0(t_{X_0}\omega)]$

$\overset{\text{Markov property}}{=} E_{0,\omega}[E_{X_n,\omega}[f_{n+1}(t_{X_1}\omega)]\, f_n(\overline{\omega}_n) \ldots f_0(\overline{\omega}_0)].$

Note that

$$E_{X_n,\omega}[f_{n+1}(t_{X_1}\omega)] = \sum_{|e|=1} p(X_n, e, \omega)\, f_{n+1}(t_{X_n+e}\,\omega) = R f_{n+1}(\overline{\omega}_n),$$

and therefore the left hand side of (1.3) equals

$$E_{0,\omega}[R\, f_{n+1}(\overline{\omega}_n)\, f_n(\overline{\omega}_n) \ldots f_0(\overline{\omega}_0)].$$

This shows the Markov property of $(\overline{\omega}_n)$ under $P_{0,\omega}$ and under $P_0$, after $\mathbb{P}$-integration. Further the distribution of $\overline{\omega}_0$ under $P_{0,\omega}$ is $\delta_\omega$ and $\mathbb{P}$ under $P_0$.     $\square$

The state space of the Markov chain of environments viewed from the particle is huge, and one might fear that this Markov chain is essentially useless. The key assumption as we shall shortly see is the existence of an invariant measure absolutely continuous with respect to $\mathbb{P}$:

(1.4)     **Key assumption:**   there exists an invariant probability $\mathbb{Q} = f\,\mathbb{P}$ for $R$
(i.e. $\int R\,h\,d\mathbb{Q} = \int h\,d\mathbb{Q}$, $h$ bounded measurable).

Intuitively this corresponds to the idea that "the static point of view and the dynamic point of view are comparable".

**Examples:**

**1)** In the case of model b) of the introduction, if one chooses:

(1.5)   $\mathbb{Q} = f(\omega)\,\mathbb{P}$, where $f(\omega) = \dfrac{1}{Z} \sum_{|e|=1} \omega(\{0, e\})$, $Z$ a normalizing constant,

then (1.4) is satisfied as follows from the stronger property:

(1.6)     $\mathbb{Q}$ is reversible for $R$ (i.e. $R$ is self-adjoint in $L^2(\mathbb{Q})$ or
for $h, g$ bounded measurable, $\int h\,R\,g\,d\,\mathbb{Q} = \int R\,h\,g\,d\mathbb{Q}$).

Indeed:

$$\int h\,R\,g\,d\mathbb{Q} \overset{(1.2)}{=} \int h \sum_{|e|=1} p(0,e,\omega)\,g\circ t_e\,d\mathbb{Q}$$

$$\overset{(1.5)}{=} \frac{1}{Z}\int h \sum_{|e|=1} \omega(\{0,e\})\,g\circ t_e\,d\mathbb{P}$$

$$\overset{\substack{\text{translation}\\ \text{invariance}}}{=} \frac{1}{Z} \sum_{|e|=1} \int (t_{-e}\,\omega)(\{0,e\})\,h\circ t_{-e}\,g\,d\mathbb{P}$$

$$= \frac{1}{Z} \sum_{|e|=1} \int \omega(\{-e,0\})\,h\circ t_{-e}\,g\,d\mathbb{P}$$

$$= \int R\,h\,g\,d\mathbb{Q}.$$

**2)** In the case of model a) of the introduction, when $d=1$, and $\mathbb{E}[\rho] < 1$, (recall $\rho(x,\omega) = \omega(x,-1)/\omega(x,1)$), one can directly check that

$$Q = f(\omega)\,\mathbb{P}, \text{ with}$$

$$(1.7) \qquad f(\omega) = \frac{1-\mathbb{E}[\rho]}{1+\mathbb{E}[\rho]}\,(1+\rho(0,\omega))(1+\rho(1,\omega)+\rho(1,\omega)\,\rho(2,\omega)$$

$$+ \rho(1,\omega)\,\rho(2,\omega)\,\rho(3,\omega)+\ldots)$$

is an invariant probability for $R$, see also Molchanov [43], p. 273. $\qquad\square$

As we shall now see the key assumption (1.4) fully determines $\mathbb{Q}$.

**Theorem 1.2.**

$$(1.8) \qquad \text{\textit{If } $\mathbb{Q}$ \textit{ satisfies} (1.4), \textit{then } $\mathbb{Q}\sim\mathbb{P}$ \textit{ and the Markov chain}}$$
$$\text{\textit{with initial law } $\mathbb{Q}$ \textit{ and transition kernel } $R$ \textit{ is ergodic.}}$$

$$(1.9) \qquad \text{\textit{There is at most one probability } $\mathbb{Q}$ \textit{ satisfying} (1.4).}$$

*Proof.* • $\mathbb{Q}\sim\mathbb{P}$: Let $E$ denote the event $\{f=0\}$. Since $\mathbb{Q}R = \mathbb{Q}$,

$$\mathbb{Q}\,R1_E = \mathbb{Q}\,1_E = \int_{\{f=0\}} f\,d\mathbb{P} = 0.$$

As a result, $R1_E = 0$, $\mathbb{P}$-a.s. on $\{f>0\}$, so that:

$$\mathbb{P}\text{-a.s.,}\ 1_E \geq R1_E = \sum_{|e|=1} p(0,e,\omega)\,1_E \circ t_e \geq \kappa \sum_{|e|=1} 1_E \circ t_e,$$

by definition of $\mathcal{P}_\kappa$ in the case of model a) and for a suitable $\kappa > 0$ for model b) (see above (1.1)). Since $1_E$ only takes the values 0 and 1, it follows that:

$$\mathbb{P}\text{-a.s.,}\ 1_E \geq 1_E \circ t_e, \text{ for } |e|=1,$$

and taking into account that $\mathbb{P}(E) = \mathbb{P}(t_e^{-1}(E))$,

$$\mathbb{P}\text{-a.s., } 1_E = 1_E \circ t_e, \text{ for } |e| = 1.$$

Composing with $t_x$, we find that:

$$\mathbb{P}\text{-a.s., } 1_E \circ t_x = 1_E \circ t_{x+e}, \text{ for all } x \in \mathbb{Z}^d \text{ and } |e| = 1,$$

from which it follows that:

(1.10)    $\mathbb{P}$-a.s., $1_E = 1_E \circ t_x$, for all $x \in \mathbb{Z}^d$.

Then $\widetilde{E} \stackrel{\text{def}}{=} \bigcap_{x \in \mathbb{Z}^d} t_x^{-1}(E)$ is invariant under every $t_y$ (i.e. $t_y^{-1}(\widetilde{E}) = \widetilde{E}$) and indistinguishable from $E$. The ergodicity of $\mathbb{P}$ under $(t_x)_{x \in \mathbb{Z}^d}$, forces that

$$\mathbb{P}(E) = \mathbb{P}(\widetilde{E}) = 0 \text{ or } 1.$$

However $\int_{E^c} f\, d\mathbb{P} = \int f\, d\mathbb{P} = 1$, so that $\mathbb{P}(E) = 0$. We thus have proved

(1.11)                              $\mathbb{Q} \sim \mathbb{P}$.

• $\mathbb{Q}$ is an ergodic invariant measure for $R$:

We let $\widetilde{\Omega} = \Omega^{\mathbf{N}}$ stand for the canonical space of $\Omega$-valued trajectories, $\widetilde{\theta}$ for the canonical shift, $\widetilde{\mathcal{B}}$ for the product $\sigma$-field, $\widetilde{P}_\omega$ for the law of the chain associated to $R$, and starting in $\omega$, $(\widetilde{\omega}_n)$ for the canonical process. We have to show that

(1.12)    for any invariant event $A$ (i.e. $A \in \widetilde{\mathcal{B}}$ and $\widetilde{\theta}^{-1}(A) = A$),
$\widetilde{P}_\mathbb{Q}(A) = 0$ or 1, where we use the notation $\widetilde{P}_\mathbb{Q} = \displaystyle\int_\Omega \widetilde{P}_\omega\, d\mathbb{Q}(\omega)$.

To prove (1.12) we define

(1.13)                          $\varphi(\omega) = \widetilde{P}_\omega(A)$.

Observe that

(1.14)    $\varphi(\widetilde{\omega}_n)$, $n \geq 0$, is a $\widetilde{P}_\mathbb{Q}$-martingale, for the canonical filtration on $\widetilde{\Omega}$.

Indeed, since $A$ is invariant,
(1.15)

$$\widetilde{E}_\mathbb{Q}[1_A \mid \widetilde{\omega}_0, \ldots, \widetilde{\omega}_n] = E_\mathbb{Q}[1_A \circ \widetilde{\theta}_n \mid \widetilde{\omega}_0, \ldots, \widetilde{\omega}_n] \stackrel{\overset{\text{Markov}}{\text{property}}}{=} \widetilde{P}_{\widetilde{\omega}_n}(A) = \varphi(\widetilde{\omega}_n), \ \widetilde{P}_\mathbb{Q}\text{-a.s.}.$$

As an application of the martingale convergence theorem and (1.15):

(1.16)                  $\varphi(\widetilde{\omega}_n) \to 1_A$, as $n \to \infty$, $\widetilde{P}_\mathbb{Q}$-a.s..

Moreover, as we shall now see

(1.17)            a) $\varphi = 1_B$, $\mathbb{Q}$-p.s., and b) $R1_B = 1_B$, $\mathbb{Q}$-p.s..

Indeed if a) was not true, then for some $a < b$ with $[a,b] \subseteq \mathbb{R}_+ \backslash \{0,1\}$, $\mathbb{Q}(\varphi \in [a,b]) > 0$. But the ergodic theorem of Birkhoff (cf. for instance the book by Durrett [19]) implies that $\widetilde{P}_\mathbb{Q}$-a.s.:

$$(1.18) \qquad \frac{1}{n} \sum_0^{n-1} 1\{\varphi(\widetilde{\omega}_k) \in [a,b]\} \to \Psi = \widetilde{E}_\mathbb{Q}[1\{\varphi(\widetilde{\omega}_0) \in [a,b]\} \,|\, \mathcal{I}],$$

where $\mathcal{I}$ stands for the $\sigma$-field of invariant events. Note that

$$\widetilde{E}_\mathbb{Q}[\Psi] = \widetilde{P}_\mathbb{Q}[\varphi(\widetilde{\omega}_0) \in [a,b]] = \mathbb{Q}[\varphi \in [a,b]] > 0,$$

and then (1.18) would contradict (1.16). Thus (1.17) a) holds. As for (1.17) b), observe that $\widetilde{P}_\mathbb{Q}$-a.s.:

$$1_B(\widetilde{\omega}_0) \overset{(1.14),(1.17)a)}{=} \widetilde{E}_\mathbb{Q}[1_B(\widetilde{\omega}_1) \,|\, \widetilde{\omega}_0] \overset{\substack{\text{Markov} \\ \text{property}}}{=} R1_B(\widetilde{\omega}_0),$$

from which (1.17) b) follows.

To prove (1.8), it will now suffice to see that:

$$(1.19) \qquad\qquad \mathbb{Q}(B) = 0 \text{ or } 1,$$

indeed $\widetilde{P}_\mathbb{Q}(A) \overset{(1.13)}{=} \int \varphi(\omega)\, d\mathbb{Q} \overset{(1.17)}{=} \mathbb{Q}(B)$.

To prove (1.19), we note that $\mathbb{Q}$-a.s. and therefore $\mathbb{P}$-a.s. by (1.11)

$$1_B \overset{(1.17)b)}{=} R1_B \geq \kappa \sum_{|e|=1} 1_B \circ t_e.$$

Now a similar argument as in the proof of (1.11) yields: $1_B = 1_B \circ t_x$, for all $x$, $\mathbb{P}$-a.s., and by ergodicity $\mathbb{P}(B) = 0$ or 1, from which (1.19) follows.

• Uniqueness of $\mathbb{Q}$:

If $g$ is a bounded measurable function on $\Omega$, it follows from Birkhoff's ergodic theorem and the fact that $\mathbb{Q}$ is invariant and ergodic for $R$, that

$$\widetilde{P}_\mathbb{Q}\text{-a.s.,}\ \frac{1}{n} \sum_0^{n-1} g(\widetilde{\omega}_k) \longrightarrow \int g\, d\mathbb{Q}, \text{ as } n \to \infty,$$

and by (1.11) $\widetilde{P}_\mathbb{P}$-a.s., as well. As a result of the proposition stating the Markov property of $(\overline{\omega}_n)$, we see that

$$E_0\Big[\frac{1}{n} \sum_0^{n-1} g(\overline{\omega}_k)\Big] \longrightarrow \int g\, d\mathbb{Q}, \text{ as } n \to \infty,$$

and the claim follows. □

We shall now close this lecture with two simple applications of the above theorem.

**Examples:**

**1)** We are in the situation of model b), and $\mathbb{Q}$ is defined as in (1.5). One defines the local drift at site $x$ in the environment $\omega$, as:

$$(1.20) \qquad d(x,\omega) = \sum_{|e|=1} p(x,e,\omega)\,e = E_{x,\omega}[X_1 - X_0].$$

Birkhoff's ergodic theorem, together with (1.8) and the proposition imply:

$$(1.21) \qquad \frac{1}{n}\sum_0^{n-1} d(X_k,\omega) = \frac{1}{n}\sum_0^{n-1} d(0,\overline{\omega}_k) \longrightarrow \int d(0,\omega)\,d\mathbb{Q},$$
$$P_{0,\omega}\text{-a.s., for }\mathbb{Q}\text{ or }\mathbb{P}\text{-a.e. }\omega.$$

On the other hand:

$$M_n = X_n - X_0 - \sum_0^{n-1} d(X_k,\omega) \text{ is a martingale with bounded increments}$$
under $P_{0,\omega}$, for $\omega \in \Omega$.

Azuma's inequality (cf. the book by Alon-Erdös-Spencer [2]) states that when $0 = Z_0, Z_1, \ldots Z_n$ is a real-valued martingale such that $|Z_{i+1} - Z_i| \le 1$, for $i \ge 0$, then

$$(1.22) \qquad \mathbb{P}[Z_n > \lambda\sqrt{n}] \le \exp\left\{-\frac{\lambda^2}{2}\right\}, \text{ for } \lambda > 0,\ n \ge 0.$$

As a result, for $|e| = 1$, $\omega \in \Omega$, $\epsilon > 0$, $n \ge 0$,

$$P_{0,\omega}[M_n \cdot e > n^{\frac{1}{2}+\epsilon}] \le \exp\left\{-\frac{n^{2\epsilon}}{8}\right\},$$

and an application of Borel Cantelli's lemma yields:

$$(1.23) \qquad \frac{M_n}{n} \to 0,\ P_{0,\omega}\text{-a.s., for }\omega \in \Omega.$$

We thus see from (1.21) and (1.23) that:

$$(1.24) \qquad P_0\text{-a.s., } \frac{X_n}{n} \longrightarrow \int d(0,\omega)\,d\mathbb{Q}.$$

On the other hand:

$$
\begin{aligned}
(1.25) \qquad \int d(0,\omega)\,d\mathbb{Q} &\overset{(1.5)}{=} \frac{1}{Z}\int \sum_{|e|=1} \omega(\{0,e\})\,e\,d\mathbb{P} \\
&= \frac{1}{Z}\sum_{i=1}^{d}\int [\omega(\{0,e_i\})\,e_i - \omega(\{0,-e_i\})\,e_i]\,d\mathbb{P} \\
&= \frac{1}{Z}\sum_{i=1}^{d}\int [\omega(\{0,e_i\}) - (t_{-e_i}\omega)(\{0,e_i\})]\,e_i\,d\mathbb{P} \\
&\overset{\substack{\text{translation}\\\text{invariance}}}{=} 0.
\end{aligned}
$$

In other words we see that model b) exhibits a null limiting velocity:

(1.26) $\qquad\qquad\qquad\qquad P_0$-a.s., $\dfrac{X_n}{n} \to 0$.

**2)** In the case of model a), when $d = 1$, and $\mathbb{E}[\rho] < 1$, defining $\mathbb{Q}$ as in (1.7), the argument above shows that

(1.27) $\qquad\qquad\qquad\qquad P_0$-a.s., $\dfrac{X_n}{n} \longrightarrow \displaystyle\int d(0,\omega)\, d\mathbb{Q}$,

(this is the way to amend the incorrect but not totally wrong argument of the introduction leading to the erroneous conclusion $\frac{X_n}{n} \to \mathbb{E}[p] - \mathbb{E}[q]$).

Observe that:

$$\int d(0,\omega)\, d\mathbb{Q} \overset{(1.7)}{=} \frac{1 - \mathbb{E}[\rho]}{1 + \mathbb{E}[\rho]}\ \mathbb{E}[d(0,\omega)(1+\rho(0,\omega))(1+\rho(1,\omega)+\rho(1,\omega)\rho(2,\omega)+\ldots)]$$

using the identity $d(0,\omega)(1 + \rho(0,\omega)) = 1 - \rho(0,\omega)$, we find

$$= \frac{1 - \mathbb{E}[\rho]}{1 + \mathbb{E}[\rho]}\ \mathbb{E}[(1 - \rho(0,\omega))(1 + \rho(1,\omega) + \ldots)] \overset{\text{independence}}{=} \frac{1 - \mathbb{E}[\rho]}{1 + \mathbb{E}[\rho]}\ .$$

We have thus found:

(1.28) $\qquad\qquad P_0$-a.s., $\dfrac{X_n}{n} \longrightarrow \dfrac{1 - \mathbb{E}[\rho]}{1 + \mathbb{E}[\rho]}\ $, when $\mathbb{E}[\rho] < 1$,

and this recovers the first case of the law of large numbers of Solomon in (0.1) of the introduction. $\qquad\qquad\qquad\qquad\qquad\qquad\qquad\qquad\qquad\qquad\qquad\square$

The point of view of the "environment viewed from the particle" has been a powerful tool in the investigation of several models like b), where the measure $\mathbb{Q}$ can be constructed "by hand". It has influenced in a crucial way developments around the central limit theorem, where some further ideas from homogenization theory enter the picture, cf. [33], [36], [46], [28], [42], [47], [6]. However, so far the "environment viewed from the particle" has had little impact in the study of multi-dimensional random walks in random environment, i.e. model a) when $d > 1$. The next lecture will present a further application of the point of view of the environment viewed from the particle, which in some sense contradicts the above remark. Indeed the lecture will revolve around a special case of model a), with $d > 1$, and $\mathbb{Q}$ will not be explicit.

## Lecture 2:

## Central Limit Theorem for Random Walks in
## Random Environment with Null Drift

We shall discuss in this lecture one instance where the point of view of the environment viewed from the particle has been efficient in the investigation of multi-dimensional random walks in random environment. This example is instructive and indicates what can be done when one does not have an explicit measure $\mathbb{Q}$ satisfying the key assumption (1.4). We want to consider the situation where $\mathbb{P}$-a.s.,

$$(2.1) \qquad d(x, \omega) \equiv 0, \text{ that is } \omega(x, e) = \omega(x, -e), \text{ for } x \in \mathbb{Z}^d, |e| = 1.$$

It is convenient to replace $\Omega = \mathcal{P}_\kappa^{\mathbb{Z}^d}$ with

$$(2.2)$$
$$\Omega_0 = \mathcal{P}_{\kappa,0}^{\mathbb{Z}^d}, \text{ where } \mathcal{P}_{\kappa,0} \text{ is the set of } (2d)\text{-vectors } (p(e))_{|e|=1} \text{ with components}$$
$$\text{in } [\kappa, 1] \text{ such that } \sum_e p(e) = 1 \text{ and } \sum_e e\,p(e) = 0,$$

and assume that

$$(2.3)$$
$$\mathbb{P} = \mu^{\otimes \mathbb{Z}^d}, \text{ where } \mu \text{ is concentrated on } \mathcal{P}_{\kappa,0}, \text{ (in fact the result we discuss}$$
$$\text{below extends to the case of a } \mathbb{P}, \text{ which is ergodic under the}$$
$$\text{action of } (t_x)_{x \in \mathbb{Z}^d}).$$

The assumption (2.1) is of course quite special in the context of model a). Observe for instance that when $d = 1$, the only case under consideration is the symmetric nearest neighbor random walk on $\mathbb{Z}$. We shall now explain the proof of the following central limit theorem of Lawler [40], (see also Papanicolaou-Varadhan [48] for the result in a continuous setting):

**Theorem 2.1.** *For $\mathbb{P}$-a.e. $\omega$, under $P_{0,\omega}$, $B^n = \frac{X_{[n\cdot]}}{\sqrt{n}}$ converges in law on $D(\mathbb{R}_+, \mathbb{R}^d)$ to a non-degenerate Brownian motion with covariance matrix which is diagonal:*

$$(2.4) \qquad\qquad A = \begin{pmatrix} a_1 & & 0 \\ & \ddots & \\ 0 & & a_d \end{pmatrix}.$$

*Sketch of Proof:* We know that for $f$ a function on $\mathbb{Z}^d$,

$$f(X_n) - f(X_0) - \sum_0^{n-1} (P_\omega f - f)(X_k), \ n \geq 0, \text{ is a } P_{0,\omega}\text{-martingale},$$

where $P_\omega f(x) = \sum_{|e|=1} p(x, e, \omega) f(x + e)$. Specializing to $f(x) = x_i$, we find

$$P_\omega f - f = 0,$$

and specializing to $f(x) = x_i x_j$,

$$(P_\omega f - f)(x) = \sum_{|e|=1} \omega(x, e)(x + e)_i (x + e)_j - x_i x_j$$

$$\overset{(2.1) \text{ or } (2.2)}{=\!=} \sum_{m=1}^d \omega(x, e_m)((x + e_m)_i (x + e_m)_j + (x - e_m)_i (x - e_m)_j - 2x_i x_j)$$

$$= \ 2\omega(x, e_i) \delta_{ij}.$$

As a result:

(2.5) $\quad X_n^i$ and $X_n^i X_n^j - 2\delta_{ij} \sum_{k=0}^{n-1} p(0, e_i, \overline{\omega}_k)$ are $P_{0,\omega}$-martingales, for

$\quad i, j \in [1, d]$, (recall $\overline{\omega}_k = t_{X_k} \omega$).

The central limit theorem for martingales (cf. Ethier-Kurtz [21], see also Durrett [20]) proves the main claim once we show that

(2.6) $\qquad$ for $\mathbb{P}$-a.e. $\omega$, $P_{0,\omega}$-a.s., $\dfrac{1}{n} \sum_0^{n-1} p(0, e_i, \overline{\omega}_k) \longrightarrow \dfrac{a_i}{2} > 0$.

Thus the main claim follows from Birkhoff's ergodic theorem and the central limit theorem for martingales, if we can construct $\mathbb{Q}$ satisfying the key assumption (1.4). Further we find

(2.7) $\qquad\qquad\qquad a_i = 2 \displaystyle\int \omega(0, e_i) \, d\mathbb{Q}(\omega).$

The idea in order to construct $\mathbb{Q}$, is to use an approximation of $\mathbb{P}$ by $\mathbb{P}_N$ which lives on $2N\,\mathbb{Z}^d$ periodic configurations, with $N \to \infty$. Then for each $N$ one can easily construct $\mathbb{Q}_N \ll \mathbb{P}_N$ invariant measure for the chain of environments viewed from the particle. The existence of $\mathbb{Q}$ satisfying (1.4), will then result from a key estimate $\sup_N \int (\frac{d\mathbb{Q}_N}{d\mathbb{P}_N})^\alpha \, d\mathbb{P}_N < \infty$, for some $\alpha > 1$.

For $N \geq 1$ we shall use the following notations:

$\Omega^N \ = \ \{\omega \in \Omega_0, \ t_x \omega = \omega, \ \forall x \in 2N\,\mathbb{Z}^d\}$, the set of $2N$-periodic configurations,

$T_N \ = \ \mathbb{Z}^d / 2N\,\mathbb{Z}^d$, and $\pi_N \colon \mathbb{Z}^d \to T_N$ the canonical projection,

$\Delta_N \ = \ \{-N, \dots, N-1\}^d$, a "fundamental domain" for $\pi_N$.

We consider an arbitrary sequence $\omega_N \in \Omega^N$, such that

$$(2.8) \qquad \mathbb{P}_N \overset{\text{def}}{=} \frac{1}{(2N)^d} \sum_{x \in \Delta_N} \delta_{t_x \omega_N} \xrightarrow{\text{weakly}} \mathbb{P}, \quad \text{as } N \to \infty,$$

(such a sequence always exists, one can for instance use the spatial ergodic theorem, cf. Krengel [37], or Dunford-Schwartz [18], and choose $\omega_N$ as the periodization of the restriction to $\Delta_N$ of a $\mathbb{P}$-typical configuration $\omega$).

For $N \geq 1$, $(X_n)$ under $P_{y,\omega_N}$ induces by projection on $T_N$ an irreducible Markov chain with finite state space, and we denote by

$$(2.9) \qquad \frac{1}{(2N)^d} \sum_{x \in T_N} \phi_N(x)\, \delta_x$$

the unique invariant probability of this Markov chain. Observe that defining $\psi_N = \phi_N \circ \pi_N$
$(2.10)$
$$\mathbb{Q}_N \overset{\text{def}}{=} \frac{1}{(2N)^d} \sum_{x \in \Delta_N} \psi_N(x)\, \delta_{t_x \omega_N} \text{ is an invariant measure for } R \text{ (cf. } (1.2)).$$

Indeed for $h$ bounded measurable on $\Omega_0$:

$$
\begin{aligned}
\mathbb{Q}_N\, Rh &= \frac{1}{(2N)^d} \sum_{x \in \Delta_N} \psi_N(x)\, Rh(t_x \omega_N) \\
&= \frac{1}{(2N)^d} \sum_{\substack{x \in \Delta_N \\ |e|=1}} \psi_N(x)\, \omega_N(x,e)\, h(t_{x+e}\, \omega_N) \\
&= \frac{1}{(2N)^d} \sum_{x \in \Delta_N} \psi_N(x)\, E_{x,\omega_N}[h(t_{X_1}\, \omega_N)] \\
&\overset{\text{invariance}}{=} \frac{1}{(2N)^d} \sum_{x \in \Delta_N} \psi_N(x)\, h(t_x\, \omega_N) \\
&= \mathbb{Q}_N\, h.
\end{aligned}
$$

The key estimate will be:

$$(2.11) \qquad \frac{1}{(2N)^d} \sum_{x \in \Delta_N} \psi_N(x)^{\frac{d}{d-1}} \leq c_1(d,\kappa)^{\frac{d}{d-1}}.$$

Let us admit (2.11) and explain how the construction of $\mathbb{Q}$ is completed. We have

$$(2.12) \qquad \mathbb{Q}_N = f_N\, \mathbb{P}_N, \quad \text{where } f_N(t_x\, \omega_N) = \frac{1}{|C_N(x)|} \sum_{y \in C_N(x)} \psi_N(y),$$
$$\text{for } C_N(x) = \{y \in \Delta_N, t_y\, \omega_N = t_x\, \omega_N\}.$$

Therefore if $C_N(x_i)$, $i = 1, \ldots, M$ form a partition of $\Delta_N$,

(2.13)

$$
\begin{aligned}
\int f_N^{\frac{d}{d-1}} \, d\mathbb{P}_N &= \sum_{i=1}^{M} f_N(t_{x_i} \omega_N)^{\frac{d}{d-1}} \frac{|C_N(x_i)|}{(2N)^d} \\
&\underset{\substack{\text{Jensen's} \\ \text{inequality}}}{\leq} \sum_{i=1}^{M} \sum_{y \in C_N(x_i)} \psi_N(y)^{\frac{d}{d-1}} \frac{1}{(2N)^d} \\
&= \frac{1}{(2N)^d} \sum_{x \in \Delta_N} \psi_N(x)^{\frac{d}{d-1}} \overset{(2.11)}{\leq} c_1^{\frac{d}{d-1}}.
\end{aligned}
$$

Using the fact that $\Omega_0$ is compact and metrizable for the product topology, we can extract a subsequence $N_k$ such that $\mathbb{Q}_{N_k} \longrightarrow \mathbb{Q}$ weakly, as $k \to \infty$. Then for $g$ continuous bounded on $\Omega_0$:

(2.14)

$$
\begin{aligned}
\left| \int g \, d\mathbb{Q} \right| &= \left| \lim_k \int g \, f_{N_k} \, d\mathbb{P}_{N_k} \right| \\
&\underset{\substack{\text{Hölder's} \\ \text{inequality}}}{\leq} \overline{\lim_k} \left( \int |g|^d \, d\mathbb{P}_{N_k} \right)^{\frac{1}{d}} \left( \int f_{N_k}^{\frac{d}{d-1}} \, d\mathbb{P}_{N_k} \right)^{\frac{d-1}{d}} \overset{(2.13)}{\leq} c_1 \|g\|_{L^d(\mathbf{P})}.
\end{aligned}
$$

As a result

(2.15)
$$
\mathbb{Q} \ll \mathbb{P} \text{ and in fact } \left\| \frac{d\mathbb{Q}}{d\mathbb{P}} \right\|_{L^{\frac{d}{d-1}}(\mathbf{P})} \leq c_1,
$$

further $\mathbb{Q}R = \mathbb{Q}$, as follows from $\mathbb{Q}_{N_k} R = \mathbb{Q}_{N_k}$, and the continuity of $R$ for the weak topology. This shows that $\mathbb{Q}$ satisfies (1.4). Incidentally observe that $\mathbb{Q}$ is unique and as a matter of fact $\mathbb{Q}_N \longrightarrow \mathbb{Q}$ weakly, (no extraction of subsequence is needed).

We are thus reduced to proving (2.11). As we now explain, (2.11) is a consequence of the following control on the resolvent:

(2.16)
$$
\sup_{x \in \Delta_N, \omega \in \Omega^N} \left| E_{x,\omega} \left[ \sum_{k \geq 0} \left( 1 - \frac{1}{N^2} \right)^k g \circ \pi_N(X_k) \right] \right| \leq c_1 N^2 \|g\|_{L^d(m_N)},
$$

with $m_N = \frac{1}{(2N)^d} \sum_{x \in T_N} \delta_x$.

Indeed,

$$
\begin{aligned}
\|\phi_N\|_{L^{\frac{d}{d-1}}(m_N)} &= \sup_{\|g\|_{L^d(m_N)} \leq 1} (\phi_N, g) \\
&\overset{(2.9)}{=} \sup_{\|g\|_{L^d(m_N)} \leq 1} \sum_{k \geq 0} \frac{1}{N^2} \left( 1 - \frac{1}{N^2} \right)^k \frac{1}{(2N)^d} \sum_{x \in \Delta_N} \psi_N(x) \, E_{x,\omega_N}[g \circ \pi_N(X_k)] \\
&\overset{(2.16)}{\leq} c_1, \text{ which proves (2.11)}.
\end{aligned}
$$

It is somewhat more convenient to transform the "periodic boundary condition" of (2.16) into a "Dirichlet boundary condition". We introduce

$$\tau \ = \ \inf\{n \geq 0, \ \|X_n - X_0\| \geq N\}, \ \ \|x\| = \sum_1^d |x_i|, \ \ D_N = \{x \in \mathbb{Z}^d, \ \|x\| < N\},$$
$$\nu \ = \ \inf\{n \geq 0, \ \|X_n\| \geq N\}.$$

**Lemma 2.2.** *(2.16) is a consequence of*

(2.17)
$$\|Q_\omega f\|_\infty \ \leq \ c_2(\kappa) N^2 \Big(\frac{1}{|D_N|} \sum_{x \in D_N} |f(x)|^d\Big)^{\frac{1}{d}}, \ \ where \ for \ \omega \in \Omega^N,$$
$$Q_\omega f(x) \ = \ E_{x,\omega} \Big[ \sum_0^{\nu-1} f(X_k) \Big], \ x \in D_N,$$
$$= \ 0, \ x \notin D_N.$$

*Proof.* Consider $\tau_0 = 0, \ \tau_1 = \tau, \ \tau_2 = \tau \circ \theta_{\tau_1} + \tau_1, \ldots, \tau_{k+1} = \tau \circ \theta_{\tau_k} + \tau_k, \ldots$ the iterates of the stopping time $\tau$. Then for $\rho = 1 - \frac{1}{N^2}$,

$$E_{x,\omega} \Big[ \sum_{k \geq 0} \rho^k \, g \circ \pi_N(X_k) \Big] = E_{x,\omega} \Big[ \sum_{m \geq 0} \sum_{\tau_m \leq k < \tau_{m+1}} \rho^k \, g \circ \pi_N(X_k) \Big]$$

and by the strong Markov property,

(2.18)
$$\leq \sum_{m \geq 0} \sup_{x \in \mathbb{Z}^d} E_{x,\omega}[\rho^\tau]^m \sup_{x \in \mathbb{Z}^d} |(Q_{t_x\omega}(g \circ \pi_N)(x + \cdot))(0)|$$
$$\overset{(2.17)}{\leq} c_2 \, N^2 \, \frac{2N}{|D_N|^{\frac{1}{d}}} \, \|g\|_{L^d(m_N)} \, (1 - \sup_x E_{x,\omega}[\rho^\tau])^{-1}.$$

Observe that for any integer $K > 0$,

(2.19)
$$E_{x,\omega}[\rho^\tau] \leq P_{x,\omega}[\tau \leq K] + \rho^K \, P_{x,\omega}[\tau > K].$$

From (2.5) and Doob's inequality (cf. Durrett [19], p. 215), when $\lambda > 0$,

$$\lambda N P_{0,t_x\omega} \Big[ \sup_{k \leq K} X_k^i \geq \lambda N \Big] \leq E_{0,t_x\omega}[(X_K^i)_+] \overset{(2.5)}{\leq} K^{\frac{1}{2}}.$$

Therefore

$$P_{x,\omega}[\tau \leq K] \leq \sum_i^d P_{0,t_x\omega} \Big[ \sup_{k \leq K} |X_k^i| \geq \frac{N}{d} \Big] \leq 2d \, \frac{d}{N} \, K^{\frac{1}{2}} \leq \frac{1}{2}, \ choosing \ K = \Big[ \frac{N^2}{16d^4} \Big].$$

Coming back to (2.19), since $\rho = 1 - \frac{1}{N^2}$, we see that uniformly in $N \geq 16d^4$, $\omega \in \Omega^N$, $\sup_x E_{x,\omega}[\rho^\tau] < 1$. This is readily extended to a uniform control for $N \geq 1$, $\omega \in \Omega^N$, and (2.16) follows from the last line of (2.18). $\qquad \square$

To prove the estimate (2.18), we use a slightly different argument from Lawler [40], who adapted the result of Krylov [38] to a discrete setting. We instead use the approach of Kuo-Trudinger, (cf. Theorem 2.1. of [39]). With no loss of generality, we assume that $f$ in (2.17) is non-negative. Observe that $u = Q_\omega f$ satisfies:

$$(2.20) \qquad \sum_{|e|=1} \omega(x,e)(u(x+e) - u(x)) = -f(x), \ x \in D_N, \ u(x) = 0, \ x \in \partial D_N.$$

We assume that $u$ is not identically zero.

For $x \in D_N$, we can consider the normal mapping of $u$ at $x$:

$$(2.21) \qquad \chi_u(x) = \{p \in \mathbb{R}^d; \ u(z) \le u(x) + p \cdot (z - x), \ \text{for} \ z \in D_N \cup \partial D_N\}.$$

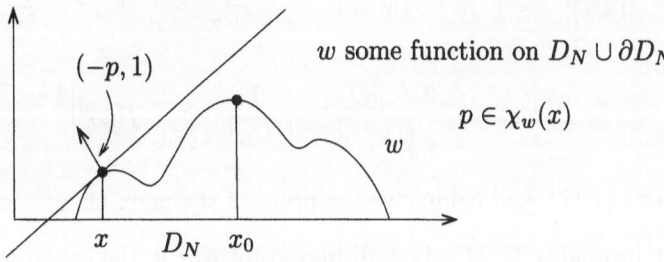

The set $\bigcup_{x \in D_N} \chi_u(x)$ is not small, for:

$$(2.22) \qquad \left\{ p \in \mathbb{R}^d, \ |p| < \frac{\max u}{2N} \right\} \subseteq \bigcup_{x \in D_N} \chi_u(x).$$

Indeed, if $|p| < \frac{\max u}{2N}$,

$$u(x_0) + p \cdot (z - x_0) > 0, \quad x \in D_N \cup \partial D_N,$$

with $u(x_0) = \max u$ and $2N = $ Euclidean diameter of $D_N \cup \partial D_N$.

Therefore if $t = \inf\{\rho \ge 0, u(x_0) + p \cdot (x - x_0) + \rho > u(x), \forall x \in D_N \cup \partial D_N\}$, $u(x) = u(x_0) + p \cdot (x - x_0) + t$, for some $x \in D_N$, and $p \in \chi_u(x)$.

We shall now see that (2.20) implies that when $p \in \chi_u(x)$

$$(2.23) \qquad p \cdot e \in \left[ u(x) - u(x - e) - \frac{f(x)}{\kappa}, u(x) - u(x - e) \right], \quad \text{for} \ |e| = 1.$$

Indeed letting $v(y) = u(x) + p \cdot (y - x)$,

$$0 = \omega(x,e)(2v(x) - v(x+e) - v(x-e)) \overset{(2.21)}{\leq} \omega(x,e)(2u(x) - u(x+e) - u(x-e))$$

writing similar inequalities with $e_i$, $i = 1, \ldots, d$ and summing up

$$\leq \sum_i^d \omega(x,e_i)(2u(x) - u(x+e_i) - u(x-e_i)) \overset{(2.20)}{=} f(x).$$

Therefore:

$$(u(x) - u(x-e)) - (u(x+e) - u(x)) \leq \frac{f(x)}{\omega(x,e)} \leq \frac{f(x)}{\kappa}.$$

But $p \in \chi_u(x)$ forces $u(x+e) - u(x) \leq p \cdot e \leq u(x) - u(x-e)$, and (2.23) follows. As a result of (2.22) and (2.23), we find

$$\omega_d \frac{(\max u)^d}{(2N)^d} = \left| B\left(0, \frac{\max u}{2N}\right) \right| \overset{(2.22)}{\leq} \sum_{x \in D_N} |\chi_u(x)| \overset{(2.23)}{\leq} \sum_{x \in D_N} \frac{f(x)^d}{\kappa^d},$$

whence

(2.24)          $$\max u \leq \frac{2N}{\omega_d^{\frac{1}{d}}} \frac{|D_N|^{\frac{1}{d}}}{\kappa} \left| \frac{1}{|D_N|} \sum_{x \in D_N} f(x)^d \right|^{\frac{1}{d}}.$$

This proves (2.17), and completes the proof of the main claim.          □

The inequality (2.24) plays an important role in the construction of $\mathbb{Q}$. It is part of a long history of estimates of similar nature in the PDE literature, related to the maximum principle of Alexandrov-Bakelman-Pucci and the Monge-Ampère equation, see for instance Kuo-Trudinger [39], Krylov [38].

# Lecture 3:

# Long Time Survival among Random Traps

We now come back to model c) of the introduction (i.e. the random traps model), and more specifically, we shall investigate the long time survival of the particle diffusing among random traps. This will illustrate a paradigm which has emerged over the recent years in the study of several models of random media, namely the preponderant role of "pockets of atypically low eigenvalues". We shall see in lecture 5 that the same paradigm appears in the context of slowdowns of random walks in random environment. We first collect some notations.

$$\Omega = \{0,1\}^{\mathbf{Z}^d}, \quad \omega(x) = 1: \text{"there is a trap in } x\text{"},$$
$$\omega(x) = 0: \text{"there is no trap in } x\text{"}.$$

$\mathbb{P} = \mu^{\otimes \mathbf{Z}^d}$      is the product of Bernoulli measures $\mu$ with success probability $1 - e^{-\nu}$ ($\nu > 0$, a fixed number).

$P_x$      stands for the canonical law of the simple random walk on $\mathbf{Z}^d$, starting in $x$.

For a given environment $\omega \in \Omega$, the set of traps is

$$(3.1) \qquad\qquad O_\omega = \{x \in \mathbf{Z}^d, \omega(x) = 1\},$$

and the entrance time of the walk in this set is denoted by:

$$(3.2) \qquad\qquad T = \inf\{n \geq 0, X_n \in O_\omega\}.$$

When discussing the long time survival of the walk among the traps, one can either consider

$$(3.3) \qquad\qquad S_\omega(n) = P_0[T > n], \text{ for } \omega \text{ a } \mathbb{P}\text{-a.s. environment},$$
this is the "quenched survival probability",

or

$$(3.4) \qquad\qquad S(n) = \mathbb{E}[S_\omega(n)] = \mathbb{P} \times P_0[T > n],$$
this is the "annealed survival probability".

The terminology "quenched" and "annealed" is inherited from metallurgy. It is a recurrent feature of the theory of random media that the respective asymptotic behaviors of "quenched" and "annealed" quantities can be substantially different. One possible way to feel the difference between (3.3) and (3.4) is to apply the spatial ergodic theorem (cf. Krengel [37] or Dunford-Schwartz [18], vol. 1), and write

$$(3.5) \qquad S(n) \stackrel{\mathbb{P}\text{-a.s.}}{=} \lim_{N \to \infty} \frac{1}{|\mathcal{T}_N|} \sum_{x \in \mathcal{T}_N} P_x[T > n], \quad \text{with } \mathcal{T}_N = [-N, N]^d .$$

This formula expresses the annealed quantity as an average in a typical environment of quenched quantities starting at points uniformly distributed over a very large box. From (3.4) one also deduces the identity:

$$(3.6) \qquad S(n) = E_0[\exp\{-\nu \, | \, X_{[0,n]}|\}] ,$$

where $|X_{[0,n]}|$ stands for the number of sites visited by the trajectory $X$ between time 0 and $n$.

In this lecture the main focus will be on the annealed quantity $S(n)$. The aforementioned eigenvalues come in the following fashion. For $U \subseteq \mathbb{Z}^d$, we introduce the sub-Markovian kernel of the walk killed when exiting $U$:

$$(3.7) \qquad P_U = 1_U \, P 1_U , \quad \text{where } P f(x) \stackrel{\text{def}}{=} \frac{1}{2d} \sum_{|x-y|=1} f(y) .$$

$P_U$ defines a self-adjoint operator on $L^2(U)$ which we shall view as the subspace of functions of $L^2(\mathbb{Z}^d)$ which vanish outside $U$. We then consider the Dirichlet form:

$$
\begin{aligned}
\mathcal{E}_U(f,f) &= \big((I - P_U)\,f, f\big), \quad f \in L^2(U), \\
&\qquad ((\cdot, \cdot) \text{ denotes the scalar product in } L^2(\mathbb{Z}^d)) \\
&= \sum_{x \in \mathbb{Z}^d} \Big(f(x) - \frac{1}{2d} \sum_{|y-x|=1} f(y)\Big) f(x) \\
(3.8) \qquad &= \frac{1}{2d} \sum_{\substack{x,y \in \mathbb{Z}^d \\ |x-y|=1}} \big(f(x) - f(y)\big) f(x) \\
&= \frac{1}{2d} \sum_{\substack{x,y \in \mathbb{Z}^d \\ |x-y|=1}} \big(f(y) - f(x)\big) f(y) \\
&= \frac{1}{4d} \sum_{\substack{x,y \in \mathbb{Z}^d \\ |x-y|=1}} \big(f(y) - f(x)\big)^2, \quad \text{(taking the half-sum of} \\
&\hspace{7cm} \text{the last two expressions).}
\end{aligned}
$$

One can then characterize $\lambda(U)$, the bottom of the spectrum of $I - P_U$ (which we shall abusively refer to as "the principal Dirichlet eigenvalue of $I - P$ in $U$", in spite of the fact that it need not be an eigenvalue when $|U| = \infty$), through the variational formula:

$$(3.9) \qquad \lambda(U) = \inf_{f \in L^2(U), f \neq 0} \frac{\mathcal{E}_U(f,f)}{(f,f)} \quad (= 1, \text{ by convention when } U = \emptyset) .$$

Note from (3.9) by choosing $f = 1_U$ that $\lambda(U) \le 1$. The quantity of main interest for our purpose is then

$$(3.10) \qquad \lambda_\omega(U) \overset{\text{def}}{=} \lambda(U \backslash O_\omega), \; \omega \in \Omega, \; U \subseteq \mathbb{Z}^d.$$

We are going to see that the large $n$ behavior of $S(n)$ is dominated by the presence of pockets $U \subseteq \mathcal{T}_n$, with atypically low $\lambda_\omega(U)$. One can give very precise versions of this type of heuristics (see for instance Chapter 4 and 7 of the book [64]). Here we shall content ourselves with:

**Proposition 3.1.**

$$(3.11) \qquad -\infty < \varliminf_n n^{-\frac{d}{d+2}} \log S(n) \le \varlimsup_n n^{-\frac{d}{d+2}} \log S(n) < 0.$$

$$(3.12) \qquad \begin{aligned} \log S(n) &\sim \log \mathbb{E}[(1 - \lambda_\omega(\mathcal{T}_n))^n] \\ &\sim \log \mathbb{E}[\exp\{-n\lambda_\omega(\mathcal{T}_n)\}], \; \text{as } \; n \to \infty. \end{aligned}$$

One can see (3.12) as an illustration of the fact that atypically low eigenvalues $\lambda_\omega(\mathcal{T}_n)$ dominate the way in which $S(n)$ tends to 0. Much more is known than (3.11), one can in fact show (cf. Donsker-Varadhan [17], Antal [3]):

**Theorem 3.2.**

$$(3.13) \qquad \begin{aligned} &\lim n^{-\frac{d}{d+2}} \log S(n) = -c_{\text{disc}}(d, \nu), \; \text{where} \\ &c_{\text{disc}}(d, \nu) = (\nu\omega_d)^{\frac{2}{d+2}} \left(\frac{d+2}{2}\right) \left(\frac{2\lambda_d}{d}\right)^{\frac{2}{d+2}}, \end{aligned}$$

with $\lambda_d$ the principal Dirichlet eigenvalue of $-\frac{1}{2d}\Delta$ in the unit ball of $\mathbb{R}^d$, and $\omega_d$ the volume of the unit ball of $\mathbb{R}^d$.

**Remark:** The quenched asymptotics is quite different from the annealed asymptotics. In particular when $\mathbb{P}$-a.s., $O_\omega^c$ only has bounded components (i.e. does not percolate, which happens for large $\nu$), it is not difficult to see that $S(n)$ decreases exponentially with $n$ for $\mathbb{P}$-a.e. $\omega$. For much more on the quenched asymptotics see Antal [3], and also Sznitman [63].

*Proof of Proposition 3.1:*

**a) (3.11): the lower bound:**

For $x \in \mathcal{T}_n$:

$$(3.14) \qquad S(n) \ge \mathbb{P} \times P_0[T > n, T_{\mathcal{T}_n - x} > n] \overset{\substack{\text{translation} \\ \text{invariance}}}{=} \mathbb{P} \times P_x[T > n, T_{\mathcal{T}_n} > n]$$

where for $U \subseteq \mathbb{Z}^d$, $T_U$ denotes the exit time from $U$:

$$(3.15) \qquad T_U = \inf\{n \ge 0, \; X_n \notin U\}.$$

Summing over $x \in \mathcal{T}_n$:

$$S(n) \geq \frac{1}{|\mathcal{T}_n|} \sum_{x \in \mathcal{T}_n} \mathbb{P} \times P_x[T_{\mathcal{T}_n \backslash O_\omega} > n] = \frac{1}{|\mathcal{T}_n|} \mathbb{E}[(1_{\mathcal{T}_n \backslash O_\omega}, P^n_{\mathcal{T}_n \backslash O_\omega} 1_{\mathcal{T}_n \backslash O_\omega})]$$

$$\underset{\text{theorem}}{\overset{\text{spectral}}{\geq}} \frac{1}{|\mathcal{T}_n|} \mathbb{E}[(1, \varphi)^2 \ (1 - \lambda_\omega(\mathcal{T}_n))^n],$$

here the expression under the expectation is understood as 0, when $\mathcal{T}_n \subseteq O_\omega$, and when $\mathcal{T}_n \backslash O_\omega \neq \emptyset$, $\varphi$ denotes an $L^2$-normalized eigenfunction of $I - P_{\mathcal{T}_n \backslash O_\omega}$ attached to the eigenvalue $\lambda_\omega(\mathcal{T}_n)$. From the last line of (3.8) and (3.9) we see that we can choose $\varphi$ non-negative, thus: $(1, \varphi) = \sum \varphi(x) \geq \sum \varphi^2(x) = 1$, since $0 \leq \varphi \leq 1$. As a result:

$$(3.16) \qquad S(n) \geq \frac{1}{|\mathcal{T}_n|} \mathbb{E}[(1 - \lambda_\omega(\mathcal{T}_n))^n].$$

If one chooses $m = [n^{\frac{1}{d+2}}]$, so that $\mathcal{T}_m \overset{(3.5)}{\subseteq} \mathcal{T}_n$, one sees that on the event $\{O_\omega \cap \mathcal{T}_m = \emptyset\}$ (i.e. no traps in $\mathcal{T}_m$),

$$(3.17) \qquad \lambda_\omega(\mathcal{T}_n) \overset{(3.9)}{\leq} \lambda_\omega(\mathcal{T}_m) \overset{(3.10)}{=} \lambda(\mathcal{T}_m) \leq \frac{c(d)}{m^2} \sim \frac{c(d)}{n^{\frac{2}{d+2}}},$$

for large $n$ (one can check the above upper bound on $\lambda(\mathcal{T}_m)$ by choosing the test function $f(x) = m^{-\frac{d}{2}} \prod_1^d \cos(\frac{\pi x_i}{2m})$, $x = (x_1, \ldots, x_d) \in \mathcal{T}_m$, in (3.9)). Thus for large $n$:

$$(3.18) \qquad S_n \geq \frac{1}{|\mathcal{T}_n|} \mathbb{P}[O_\omega \cap \mathcal{T}_m = \emptyset] \left(1 - \frac{c(d)}{m^2}\right)^n \geq \frac{1}{|\mathcal{T}_n|} e^{-\nu|\mathcal{T}_m| - 2\frac{c(d)}{m^2} n}$$

as a result $\varliminf n^{-\frac{d}{d+2}} \log S(n) > -\infty$, and the leftmost inequality of (3.11) follows.

## b) (3.11): the upper bound:

Note that $P_0$-a.s., $T_{\mathcal{T}_n} > n$, thus for $\omega \in \Omega$:

$$(3.19) \qquad \begin{aligned} P_0[T > n] &= P_0[T_{\mathcal{T}_n \backslash O_\omega} > n] = (\delta_0, P^n_{\mathcal{T}_n \backslash O_\omega} 1_{\mathcal{T}_n \backslash O_\omega}) \\ &\leq \|\delta_0\|_{L^2} (1 - \lambda_\omega(\mathcal{T}_n))^n \|1_{\mathcal{T}_n}\|_{L^2} \leq c(d) n^{\frac{d}{2}} (1 - \lambda_\omega(\mathcal{T}_n))^n. \end{aligned}$$

Taking $\mathbb{P}$-expectations:

$$(3.20) \qquad S(n) \leq c(d) n^{\frac{d}{2}} \mathbb{E}[(1 - \lambda_\omega(\mathcal{T}_n))^n], \ n \geq 1.$$

To finish the proof of (3.11), it thus suffices to show that:

$$(3.21) \qquad \varlimsup n^{-\frac{d}{d+2}} \log \mathbb{E}[(1 - \lambda_\omega(\mathcal{T}_n))^n] < 0,$$

which in turns will follow from the key estimate:

$$(3.22) \qquad \exists c_0(d, \nu) > 0, \ \varlimsup_n n^{-\frac{d}{d+2}} \log \mathbb{P}[\lambda_\omega(\mathcal{T}_n) \leq c_0 n^{-\frac{2}{d+2}}] < 0,$$

which controls the probability of occurrence of small eigenvalues. To prove (3.22) we shall employ a strategy analogous to Kirsch-Martinelli [35] in their study of the Lifshitz tail of the density of states. Here again it is possible to obtain substantially finer results than (3.22), with the help of the "method of enlargement of obstacles", cf. Chapter 4 and 7 of the book [64], see also Antal [4], and Ben Arous-Ramirez [5].

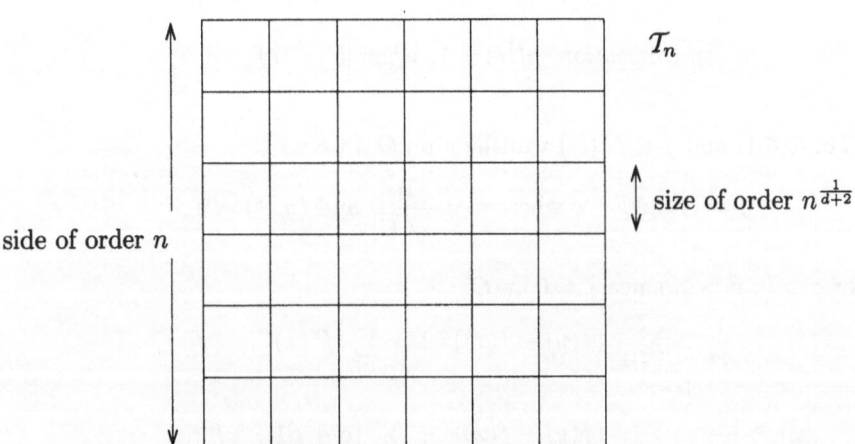

side of order $n$

$\mathcal{T}_n$

size of order $n^{\frac{1}{d+2}}$

We "chop" $\mathcal{T}_n$ into disjoint boxes $B_i$, $i \in [1, M(n)]$ with side length between $m$ and $2m$, so that (a very rough bound) $M(n) \leq c(d)n^d$. Then for $\omega \in \Omega$, and $f \in L^2(\mathcal{T}_n \backslash O_\omega)$,

$$\mathcal{E}_{\mathcal{T}_n \backslash O\omega}(f, f) \overset{(3.8)}{=} \frac{1}{4d} \sum_{\substack{x,y \in \mathbf{Z}^d \\ |x-y|=1}} (f(y) - f(x))^2 \geq \sum_{i=1}^{M(n)} \frac{1}{4d} \sum_{\substack{x,y \in B_i \\ |x-y|=1}} (f(y) - f(x))^2$$

(this is a type of "Dirichlet-Neumann bracketing" argument, cf. Reed-Simon [52], vol. IV), whence

$$\lambda_\omega(\mathcal{T}_n) \geq \inf_{i \in [1, M(n)]} \tilde{\lambda}_\omega(B_i), \quad \text{where for } U \subseteq \mathbf{Z}^d:$$

(3.23) $\tilde{\lambda}_\omega(U) = \inf \left\{ \frac{1}{4d} \sum_{\substack{x,y \in U \\ |x-y|=1}} (f(y) - f(x))^2, \ f = 0 \text{ on } O_\omega, \ \sum_{x \in U} f^2(x) = 1 \right\}.$

We now have a lower bound for $\tilde{\lambda}_\omega(U)$ which involves the size of $O_\omega$ in $U$:

**Lemma 3.3.** *(Thirring's inequality)*

*For $U \subset \mathbb{Z}^d$, finite non-empty, and for $\omega \in \Omega$:*

$$\tilde{\lambda}_\omega(U) \geq \mu(U) \frac{|O_\omega \cap U|}{|U|}, \text{ with}$$

$$(3.24) \quad \mu(U) = \inf\left\{\frac{1}{4d} \sum_{\substack{x,y \in U \\ |x-y|=1}} (f(y) - f(x))^2, \sum_{x \in U} f(x) = 0, \sum_{x \in U} f^2(x) = 1\right\}$$

*(by convention, $\mu(U) = 1$, when $|U| = 1$).*

*Proof.* For $\omega \in \Omega$ and $f \in L^2(U)$ vanishing on $O_\omega$, we write:

$$f = (f, \psi)\psi + g \text{ where } \psi = \frac{1_U}{\sqrt{|U|}} \text{ and } (g, \psi) = 0.$$

Then for $\alpha > 0$, $\beta > 0$, since $f = 0$ on $O_\omega$:

$$\frac{1}{4d} \sum_{\substack{x,y \in U \\ |x-y|=1}} (f(y) - f(x))^2 + \alpha \sum_{x \in U} f^2(x)$$

$$(3.25) \quad \geq \frac{1}{4d} \sum_{\substack{x,y \in U \\ |x-y|=1}} (f(y) - f(x))^2 + \sum_{x \in U} (\alpha + \beta 1_{O_\omega}) f^2(x)$$

$$= \frac{1}{4d} \sum_{\substack{x,y \in U \\ |x-y|=1}} (g(y) - g(x))^2 + \sum_{x \in U} (\alpha + \beta 1_{O_\omega}) f^2(x)$$

using now the definition of $\mu(U)$ to write a lower bound on the first term of the last line of (3.25) and Cauchy-Schwarz's inequality to derive a lower bound on the last term, we find:

$$\geq \mu(U) \sum_{x \in U} g^2(x) + \frac{1}{\sum_U \frac{1}{\alpha + \beta 1_{O_\omega}} \psi^2} (f, \psi)^2$$

since $\|f\|_{L^2}^2 = \|g\|_{L^2}^2 + (f, \psi)^2$, from the definition of $\tilde{\lambda}_\omega(U)$ we find:

$$(3.26) \quad \tilde{\lambda}_\omega(U) + \alpha \geq \min\left(\mu(U), \frac{1}{\sum_U \frac{1}{\alpha + \beta 1_{O_\omega}} \psi^2}\right).$$

Since $\psi^2 = \frac{1}{|U|} 1_U$, letting $\beta \to \infty$, we find:

$$\tilde{\lambda}_\omega(U) \geq \min\left(\mu(U), \alpha \frac{|U|}{|U \setminus O_\omega|}\right) - \alpha,$$

optimizing over $\alpha$, we choose $\alpha = \mu(U) \frac{|U \setminus O_\omega|}{|U|}$ if this latter quantity is not zero, otherwise we let $\alpha$ tend to 0, the claim (3.24) follows.          $\square$

We can now proceed with the proof of (3.22). For $c > 0$,

$$
\begin{aligned}
\mathbb{P}[\lambda_\omega(\mathcal{T}_n) \leq cm^{-2}] &\overset{(3.23)}{\leq} M(n) \sup_{i \in [1, M(n)]} \mathbb{P}[\widetilde{\lambda}_\omega(B_i) \leq cm^{-2}] \\
&\overset{(3.24)}{\leq} M(n) \sup_{i \in [1, M(x)]} \mathbb{P}\left[\mu(B_i) \frac{|O_\omega \cap B_i|}{|B_i|} \leq cm^{-2}\right].
\end{aligned}
$$

(3.27)

It is not hard to see, using the fact that the $B_i$ are products of intervals, that for $c_1(d) > 0$, and $n \geq 1$, $i \in [1, M(n)]$, $\mu(B_i) \geq \frac{c_1}{m^2}$, (recall $m = [n^{\frac{1}{d+2}}]$), (see Lemma 2.2.1 of Saloff-Coste [58]). Therefore the left hand side of (3.27) is smaller than

$$
M(n) \sup_{i \in [1, M(n)]} \mathbb{P}\left[\frac{|O_\omega \cap B_i|}{|B_i|} \leq \frac{c}{c_1}\right].
$$

However under $\mathbb{P}$, $|O_\omega \cap B_i|$ is a binomial variable with parameters $|B_i|$ and $1 - e^{-\nu}$, moreover $M(n) \leq c(d)\, n^d$. If we choose $c = c_0 < c_1(1 - e^{-\nu})$, it follows from a Cramer-type estimate that the above quantity for large $n$ is smaller than $\exp\{-\text{const}\, m^d\}$, and (3.22) follows.

**c)** *Proof of (3.12):*

In view of the very quantitative lower bound (3.16) and upper bound (3.19), together with (3.11), we see that

$$
\log S(n) \sim \log \mathbb{E}[(1 - \lambda_\omega(\mathcal{T}_n))^n].
$$

The fact that $\log \mathbb{E}[(1 - \lambda_\omega(\mathcal{T}_n))^n] \sim \log \mathbb{E}[\exp\{-n\lambda_\omega(\mathcal{T}_n)\}]$ is also straightforward. $\qquad\square$

The proof of the crucial bound (3.22) shows that on the event $\{\lambda_\omega(\mathcal{T}_n) \leq c_0\, n^{-\frac{2}{d+2}}\}$, a "hole" appears in scale $n^{\frac{1}{d+2}}$ in the trap configuration. One can go much further with the "method of enlargement of obstacles" and see that conditionally on the occurrence of an atypically low eigenvalue in a box, with high probability there is formation of a spherical hole, see for instance Chapter 4 §4 A in [64], and also Proposition 2 of Povel [51]. In the case of a Brownian motion moving among obstacles which are the translates of a fixed nonpolar compact set (for instance a closed ball) to the points of a Poisson point process with intensity $\nu > 0$, one defines in analogy with (3.4):

(3.28)
$$
S(t) = \mathbb{P} \times P_0[T > t],
$$

where $T$ stands for the entrance time in the union of obstacles, $\mathbb{P}$ for the law of the Poisson point process and $P_0$ for the Wiener measure. One has the analogue of (3.13), with a constant $c_{\text{cont}}(d, \nu)$ similar to $c_{\text{disc}}(d, \nu)$, except for the fact that $-\frac{1}{2d}\Delta$ is replaced by $-\frac{1}{2}\Delta$. As a matter of fact:

(3.29)
$$
c_{\text{cont}}(d, \nu) = \inf_{U \text{ open}} \{\nu|U| + \lambda_{-\frac{1}{2}\Delta}(U)\},
$$

with $\lambda_{-\frac{1}{2}\Delta}(U)$ the principal Dirichlet eigenvalue of $-\frac{1}{2}\Delta$ in $U$, and the infimum is attained for balls of radius

$$(3.30) \qquad R_0(d,\nu) = \left(\frac{2\lambda_d}{d\nu\,\omega_d}\right)^{\frac{1}{d+2}},$$

(see for instance Chapter 4 §5 of [64]). Further one has the following "confinement property", which was only very recently proved when $d \geq 3$, in Povel [51]:

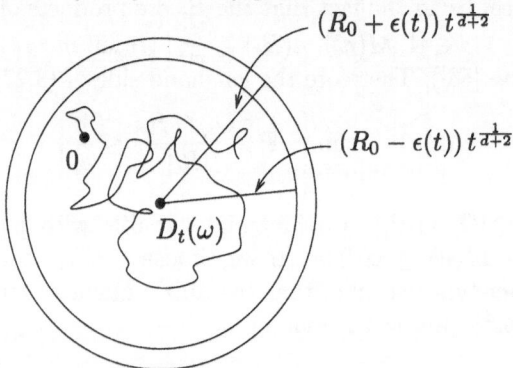

**Theorem 3.4.** *(confinement property) There exist* $\epsilon(t) \to 0$, $D_t(\omega) \in B(0,(R_0 + \epsilon(t))\,t^{\frac{1}{d+2}})$ *such that with* $\mathbb{P} \times P_0[\,\cdot\,|T > t]$*-probability tending to 1 as* $t \to \infty$:

(3.31) *the particle does not exit* $B(D_t, (R_0 + \epsilon(t))\,t^{\frac{1}{d+2}})$ *during* $[0,t]$,

(3.32) *there are no obstacle* $(d=2)$, *few obstacles (in the capacity sense,* $d \geq 3$)
$\qquad$ *inside* $B(D_t, (R_0 - \epsilon(t))\,t^{\frac{1}{d+2}})$.

The analogue when $d = 1$, can be found in Schmock [59]. The second statement (3.32) when $d \geq 3$ is not explicitly in [51] but can be easily deduced from Theorem 3.2.3 of [64] and the results of [51]. The result when $d = 2$ is proved in Sznitman [62], and in the case of $\mathbb{Z}^2$ by Bolthausen [7]. The case of dimension $d \geq 3$ remained open for quite some time. One of the reasons is that a simply connected domain of $\mathbb{R}^d$ can have a boundary with $(d-1)$-surface measure close to that of a ball with same volume, but still be far from any ball in the Hausdorff sense.

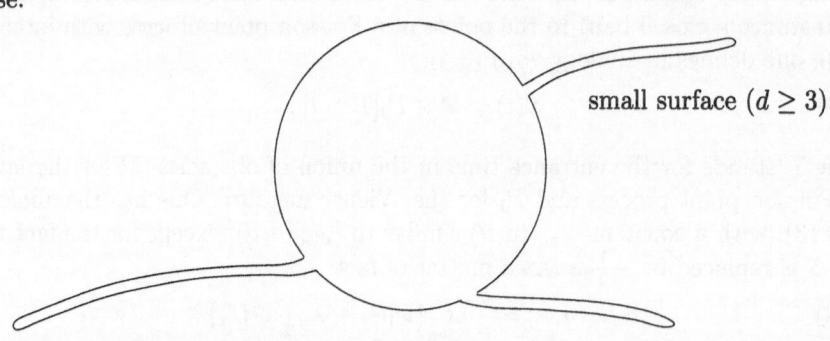

small surface $(d \geq 3)$

The isoperimetric inequality of R.R. Hall [25] replaces in the higher-dimensional proof of the confinement property by Povel [51], the role of Bonnesen's isoperimetric inequality in the two-dimensional proof [62]. It provides a control on the Lebesgue measure of the symmetric difference of the domain with some appropriate ball.

# Lecture 4:

# Multi-Dimensional Random Walks in Random Environment

We now return to model a) of the Introduction. Until recently relatively few works have discussed the behavior of multi-dimensional random walks in random environment: Kalikow [29] establishes certain $0-1$ laws and a sufficient criterion for transience, Lawler [40] shows a central limit theorem in the special situation where $d(x,\omega) \equiv 0$, (see Lecture 2), and Bricmont-Kupiainen [9] derives a central limit theorem in the isotropic case, when $d \geq 3$, for small perturbations of the simple random walk on $\mathbb{Z}^d$. More recently, some new results have been obtained, cf. Zerner [73], Sznitman-Zerner [68], Sznitman [65], [66], [67]. We shall describe some of them in this and the next lecture. As we shall see, especially in the next lecture, another example of "preponderant role of atypical pockets of low local eigenvalues" emerges in this context.

When $d > 1$, the random walk in random environment is an irreversible Markov chain under the quenched measure. Indeed Kolmogorov's criterion (cf. the book of Durrett [19]) characterizes the existence of a reversible measure (i.e. $m(x) \geq 0$, with $m(x)\, p(x,y) = m(y)\, p(y,x)$, for all $x,y$) of a Markov chain with transition probability $p(x,y)$ on the at most denumerable set $E$, through the two requirements:

$$\text{i)} \quad p(x,y) > 0 \Longrightarrow p(y,x) > 0,\ x,y \in E,$$

(4.1)    ii)   "$\log \dfrac{p(x,y)}{p(y,x)}$ is a closed form", i.e. for any loop $x_0, x_1, \ldots, x_m = x_0$,

$$\text{with}\ \prod_{i=0}^{m-1} p(x_i, x_{i+1}) > 0,\ \sum_{i=0}^{m-1} \log \frac{p(x_i, x_{i+1})}{p(x_{i+1}, x_i)} = 0\,.$$

For random walks in random environment, when $d > 1$, we can choose a simple loop: $x_0 = 0$, $x_1 = e$, $x_2 = e + e'$, $x_3 = e'$, $x_4 = 0$, with $e, e'$ of length 1, and orthogonal. Observe that

$$(4.2) \qquad C(\omega) = \sum_{i=0}^{3} \log \frac{\omega(x_i, x_{i+1} - x_i)}{\omega(x_{i+1}, x_i - x_{i+1})} = \log \frac{\omega(0,e)}{\omega(0,e')} + h(\omega)$$

where the last two random variables are independent under $\mathbb{P}$. If $\mu$ (the law of $\omega(0, \cdot)$) is not degenerate, we can choose $e$ and $e'$ so that $\operatorname{var}(C(\omega)) > 0$. As a result of Borel-Cantelli's lemma, we see that

$$\mathbb{P}\text{-a.s.},\ C(t_x \omega) \neq 0,\ \text{for infinitely many } x\,.$$

This shows the generic irreversibility of the random walk in random environment. In fact the model is genuinely irreversible in the sense that one has so far little understanding of invariant measures of the quenched Markov chain (possibly restricted in some way to a large finite box).

**Traps:**

This is maybe the most important effect for random walks in random environment. Heuristically, traps are "pockets where the walk may spend a comparatively long time with an atypically high probability". We shall now give a more quantitative description in spectral terms. For $U \subset \mathbb{Z}^d$, non-empty, we consider the kernel of the chain in the environment $\omega \in \Omega$, killed when exiting $U$ (compare with (3.7)):

$$(4.3) \qquad R_{1,\omega}^U = 1_U \, P_\omega \, 1_U, \quad \text{where } P_\omega f(x) = \sum_{|e|=1} \omega(x,e) \, f(x+e) \,.$$

Using the notation $R_{n,\omega}^U = (R_{1,\omega}^U)^n$, we have for bounded $f : \mathbb{Z}^d \to \mathbb{R}$,

$$(4.4) \qquad R_{n,\omega}^U \, f(x) = E_{x,\omega}[f(X_n), \, T_U > n], \, x \in \mathbb{Z}^d \,,$$

moreover the norm of $R_{n,\omega}^U$ from $L^\infty(U)$ in $L^\infty(U)$ is:

$$(4.5) \qquad \|R_{n,\omega}^U\|_{\infty,\infty} = \sup_x P_{x,\omega}[T_U > n] \,.$$

From the obvious submultiplicativity, we can define by superadditivity

$$
\begin{aligned}
\lambda_\omega(U) &\stackrel{\text{def}}{=} \lim_n \, -\frac{1}{n} \, \log \|R_{n,\omega}^U\|_{\infty,\infty} \in [0,\infty] \\
&= \sup_n \, -\frac{1}{n} \, \log \|R_{n,\omega}^U\|_{\infty,\infty} \,.
\end{aligned}
$$

(4.6)

In other words, $e^{-\lambda_\omega(U)}$ is the spectral radius of $R_{1,\omega}^U$. The number $\lambda_\omega(U)$ plays an analogous role to the principal Dirichlet eigenvalue defined in (3.10). It enables to quantify the strength of the trap created by $\omega$ in $U$: the smaller $\lambda_\omega(U)$ the stronger the trap. As a result of the second line of (4.6), we have a quantitative (i.e. not exclusively asymptotic) lower bound

$$(4.7) \qquad \sup_x \, P_{x,\omega}[T_U > n] \geq \exp\{-n\,\lambda_\omega(U)\}, \, n \geq 0, \, \omega \in \Omega \,.$$

However $e^{-n\lambda_\omega(U)}$ need not provide a quantitative upper bound on the decay of $\sup_x P_{x,\omega}[T_U > n]$. As a toy-example consider the degenerate situation where $d = 1$, $\omega(x,1) \equiv 1$, and $U = \{0, \ldots, n\}$.

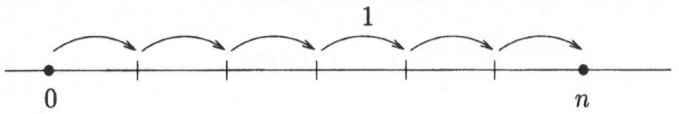

Then $\lambda_\omega(U) = \infty$ (one cannot survive for a long time in $U$), but $P_{0,\omega}[T_U > n] = 1$ is not dominated by $\exp\{-n\,\lambda_\omega(U)\}$. We are in presence of a non-self-adjoint situation and spectral theory is not all too quantitative, (compare with (3.19)).

The nature of possible traps depends on the law of the local drift $d(x,\omega)$ at a point. Introduce

(4.8)     $K_0 =$ the convex hull of the law of $d(0,\omega)$ under $\mathbb{P}$ (in fact under $\mu$).

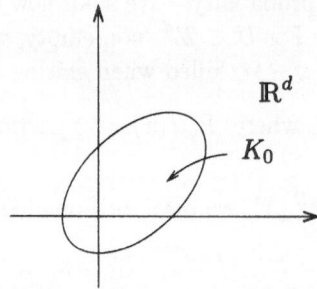

We let $B_L$ stand for the ball $B(0,L)$ in $\mathbb{Z}^d$. How small $\lambda_\omega(B_L)$ can get, is the object of the next

**Proposition 4.1.** *There exist* $c_1(d,\mu)$, $c_2(d,\mu) > 0$, *such that when:*

(4.9)     $0 \notin K_0$ *(the non-nestling*     $\mathbb{P}$-*a.s.*, $c_2 \leq \lambda_\omega(B_L) \leq c_1$, $L > 1$,
                    *case)*,

(4.10)     $0 \in \partial K_0$ *(the marginal*     $\mathbb{P}$-*a.s.*, $\dfrac{c_2}{L^2} \leq \lambda_\omega(B_L)$, $L > 1$, *and*

                    *nestling case)*,     $\mathbb{P}\Big[\lambda_\omega(B_L) \leq \dfrac{c_1}{L^2}\Big] > 0$, $L > 1$,

(4.11)     $0 \in \overset{\circ}{K}_0$ *(the plain*     $\mathbb{P}$-*a.s.*, $\exp\{-c_2\,L\} \leq \lambda_\omega(B_L)$, $L > 1$, *and*

                    *nestling case)*,     $\mathbb{P}[\lambda_\omega(B_L) \leq \exp\{-c_1 L\}] > 0$, $L$ *large*.

*Sketch of proof:* We shall merely discuss the proof of (4.11), for more details, we refer to Sznitman [67].

**a) The lower bound:**

From the definition of $\mathcal{P}_\kappa$ (cf. Lecture 1), $\mathbb{P}$-a.s., for $L > 1$,

$\inf\limits_{B_L} P_{x,\omega}[T_{B_L} \leq L + 1] \geq \kappa^{L+1}$, and therefore:

$$\sup\limits_{B_L} P_{x,\omega}[T_{B_L} > n] \overset{\overset{\text{Markov property}}{}}{\leq} (1 - \kappa^{L+1})^{[\frac{n}{L+1}]} \leq \dfrac{1}{1-\kappa}\,\exp\Big\{-\dfrac{n}{L+1}\,\kappa^{L+1}\Big\}.$$

Thus from (4.5) and (4.6), for $\omega \in \Omega$, $L > 1$:

$$\lambda_\omega(B_L) \geq e^{-\gamma L}, \quad \text{with} \quad \gamma = \log \frac{1}{\kappa} + \sup_{v \geq 1} \frac{\log\{(1 + v)/\kappa\}}{v}.$$

**b) The upper bound:**

We shall now create a "trap". We first observe that

$$w \in S^{d-1} \longrightarrow \mathbb{E}[(d(0, \omega) \cdot w)_-]$$

is a continuous function which does not vanish since $0 \in \overset{\circ}{K}$. Thus we define $\gamma_1(d, \mu) > 0$:

$$\min_{S^{d-1}} \mathbb{E}[(d(0, \omega) \cdot w)_-] = 2\gamma_1.$$

Using now the inequality:

$$P[X > \tfrac{1}{2} E[X]] \geq \tfrac{1}{4} \frac{E[X]^2}{E[X^2]}, \quad \text{for } X \text{ a non-negative random variable},$$

we see that for $w \in S^{d-1}$

$$(4.12) \qquad \mathbb{P}[(d(0, \omega) \cdot w)_- \geq \gamma_1] \geq \frac{1}{4} \frac{\mathbb{E}[(d(0, \omega) \cdot w)_-]^2}{\mathbb{E}[(d(0, \omega) \cdot w)_-^2]} \geq \gamma_1^2 > 0.$$

We can now introduce the "trapping event"

$$(4.13) \qquad T_L = \left\{ \omega : \forall x \in B_L \backslash \{0\}, \ d(x, \omega) \cdot \frac{x}{|x|} \leq -\gamma_1 \right\}$$

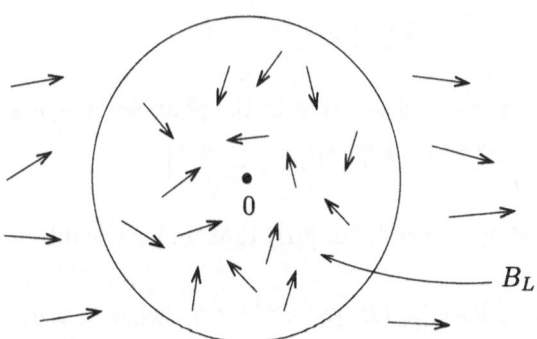

**Lemma 4.2.** *There exist $\gamma_2(d, \mu)$, $\gamma_3(d, \mu) > 0$, such that for $L > \gamma_2$, on the event $T_L$:*

$$(4.14) \qquad E_{x,\omega}[\exp\{\gamma_3 |X_1|\}] \leq \exp\{\gamma_3 |x|\}, \ x \in B_L \backslash B_{\gamma_2}.$$

*Proof.*

$$\sum_{|e|=1} \omega(x,e) e^{\gamma_3(|x+e|-|x|)} \stackrel{\text{def}}{=} 1 + \gamma_3 \, d(x,\omega) \cdot \frac{x}{|x|} + \Delta, \quad \text{and}$$

$$|\Delta| \le \sum_{|e|=1} \omega(x,e) \left| \exp\left\{ \gamma_3 \frac{2x \cdot e + 1}{|x+e| + |x|} \right\} - 1 - \gamma_3 \, e \cdot \frac{x}{|x|} \right|.$$

Note that $\frac{2x \cdot e + 1}{|x+e|+|x|} \le 3$ and for $|u| \le 1$ and a suitable $\gamma > 0$, $|e^u - 1 - u| \le \gamma u^2$. Therefore for a suitable $\gamma' > 0$, when $\gamma_3 \le \frac{1}{3}$ and $|x| \ge 1$:

$$|\Delta| \le \gamma \, 9\gamma_3^2 + \gamma' \frac{\gamma_3}{|x|}.$$

Thus choosing $\gamma_3$ small enough and $|x| \ge \gamma_2$ large enough, we see that on $\mathcal{T}_L$, (4.14) holds. □

Using a supermartingale argument, we deduce from (4.14) that for $L > \gamma_2$, on $\mathcal{T}_L$:

$$E_{x,\omega}\left[ \exp\left\{ \gamma_3 \, |X_{T_{B_L \setminus B_{\gamma_2}}}| \right\} \right] \le \exp\{\gamma_3 |x|\}, \quad x \in B_L \setminus B_{\gamma_2}.$$

Thus for large $L$, on $\mathcal{T}_L$:

$$P_{0,\omega}[T_{B_L} > n] \ge \left( \inf_{\partial B_{\gamma_2}} P_{x,\omega}[T_{B_L} > H_{B_{\gamma_2}}] \right)^n \ge \left( 1 - e^{-\frac{\gamma_3}{2} L} \right)^n, \quad n \ge 0,$$

provided

(4.15)    $H_U = \inf\{n \ge 0, \ X_n \in U\}$, denotes the entrance time in $U$, for $U \subset \mathbb{Z}^d$.

Therefore for large $L$, on $\mathcal{T}_L$:

$$(4.16) \qquad \lambda_\omega(B_L) \le \overline{\lim} - \frac{1}{n} \log P_{0,\omega}[T_{B_L} > n] \le \log \frac{1}{1 - e^{\frac{-\gamma_3}{2} L}} \le e^{-\frac{\gamma_3}{3} L},$$

Note incidentally that:

$$(4.17) \qquad \mathbb{P}[\mathcal{T}_L] \ge \gamma_1^{2|B_L|} > 0.$$ □

**Remark:** The above argument shows that in the plain nestling case

$$(4.18) \qquad \varliminf_L L^{-d} \log \mathbb{P}[\lambda_\omega(B_L) \le e^{-\frac{\gamma_3}{2} L}] > -\infty.$$

It can be shown (cf. Proposition 4.7 of [67]) that under Condition (T), see (4.27) below, one also has:

$$(4.19) \qquad \varlimsup_L L^{-d} \log \mathbb{P}[\lambda_\omega(B_L) \le e^{-cL}] < 0, \quad \text{when } c > 0.$$ □

We have seen in Lecture 3, that pockets of low principal Dirichlet eigenvalue play an important role in the long time survival of the simple random walk among random traps. Somewhat in a similar spirit, traps provide a way to slowdown random walks in random environment. This fact is already apparent in the next

**Proposition 4.3.** *For $L > 0$, $n \geq 0$:*

(4.20) $\qquad P_0[|X_n| < 2L] \geq P_0[T_{B_{2L}} > n] \geq \dfrac{1}{|B_L|} E[\exp\{-n\,\lambda_\omega(B_L)\}]$

(*compare with* (3.16)).
*Proof.*

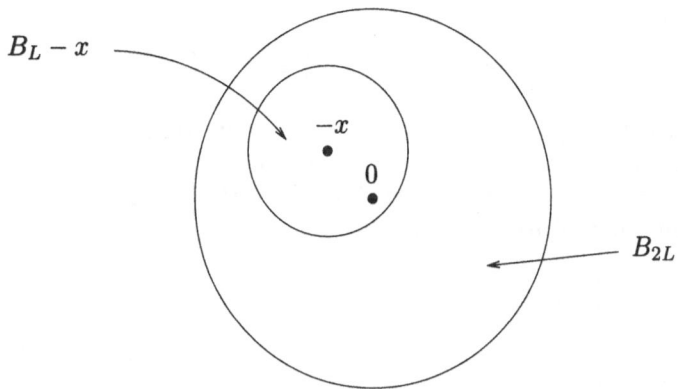

For $x \in B_L$:

$$P_0[T_{B_{2L}} > n] \geq P_0[T_{B_L - x} > n] \overset{\substack{\text{translation} \\ \text{invariance}}}{=} P_x[T_{B_L} > n]\,,$$

whence summing over $x \in B_L$,

$$P_0[T_{B_{2L}} > n] \;\geq\; \frac{1}{|B_L|} \sum_{x \in B_L} P_x[T_{B_L} > n] = \frac{1}{|B_L|} \,\mathbb{E}\Big[\sum_{x \in B_L} P_{x,\omega}[T_{B_L} > n]\Big]$$

$$\overset{(4.7)}{\geq} \frac{1}{|B_L|}\,\mathbb{E}[\exp\{-n\lambda_\omega(B_L)\}]\,.$$

$\hfill\square$

**The condition (T):**

We shall further explore the connection between traps and slowdowns of random walks in random environment. It will be convenient to this end to define a class of random walks in random environment which display a "ballistic behavior", when $d \geq 2$. Let us incidentally mention that questions concerning recurrence and transience of random walks in random environment are presently very poorly understood. For instance, S. Kalikow who was a student of H. Kesten has shown in his thesis [29] that in general, for the model a), when $\ell \in S^{d-1}$:

(4.21) $\qquad P_0[\{\lim_n X_n \cdot \ell = \infty\} \cup \{\lim_n X_n \cdot \ell = -\infty\}] = 0 \ \text{ or } \ 1\,,$

however it was not known in general when $d \geq 2$, whether or not $P_0[\lim X_n \cdot \ell = \infty]$ satisfies a zero-one law. This $0-1$ law has very recently been proved when $d = 2$, by Zerner and Merkl [74]. Omer Adelman has also announced results on this question.

We are going to introduce a renewal structure, and need to this end some notations. For $\ell \in S^{d-1}$, $u \in \mathbb{R}$, we define

(4.22)
$$T_u^\ell = \inf\{n \geq 0, \; X_n \cdot \ell \geq u\}, \quad \widetilde{T}_u^\ell = \inf\{n \geq 0, \; X_n \cdot \ell \leq u\}$$
$$D^\ell = \inf\{n \geq 0, \; X_n \cdot \ell < X_0 \cdot \ell\}.$$

We now suppose that $\ell \in S^{d-1}$ is such that:

(4.23)
$$P_0[\lim_n X_n \cdot \ell = \infty] = 1.$$

We are going to define for an arbitrary $a > 0$, a random variable $\tau_1$, $P_0$-a.s. finite which is "the first time where $X_n \cdot \ell$ goes by an amount at least $a$ above its previous local maxima, and never goes below this level from then on", (this variable will not be a stopping time relative to the natural filtration of $X_n$).

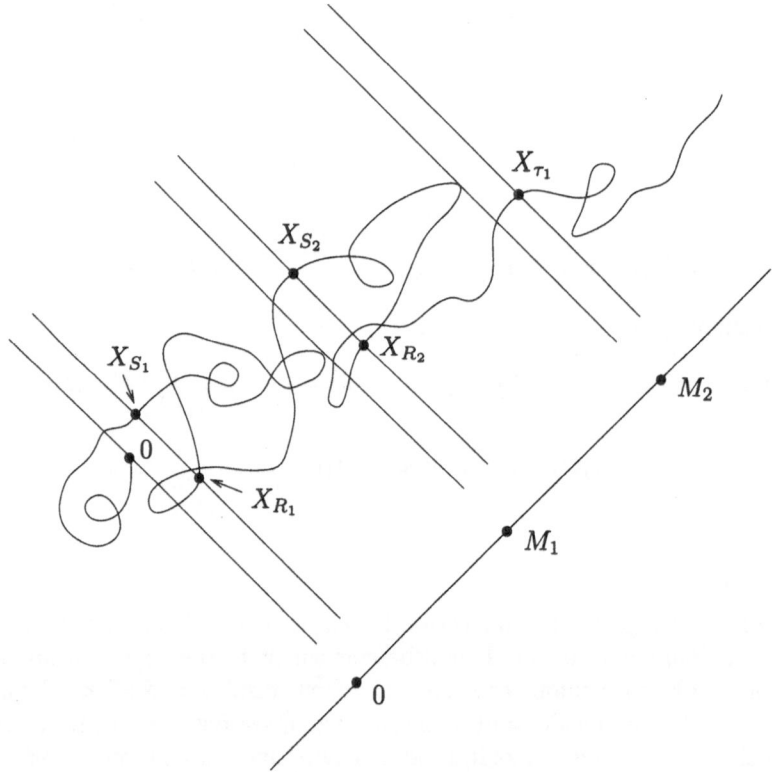

More precisely:

$$S_0 = 0, \; M_0 = X_0 \cdot \ell$$
$$S_1 = T_{M_0+a}^\ell \leq \infty, \qquad R_1 = D^\ell \circ \theta_{S_1} + S_1 \leq \infty,$$
$$M_1 = \sup\{X_n \cdot \ell, \; 0 \leq n \leq R_1\} \leq \infty,$$

($\theta$ denotes the shift on the canonical space of trajectories) and by induction:

$$S_{k+1} = T^\ell_{M_k+a} \leq \infty, \quad R_{k+1} = D^\ell \circ \theta_{S_{k+1}} + S_{k+1} \leq \infty,$$
$$M_{k+1} = \sup\{X_n \cdot \ell, \, 0 \leq n \leq R_{k+1}\}.$$

Under (4.23), it is not hard to show (cf. Proposition 1.2 of Sznitman-Zerner [68]),

(4.24)
$$P_0[D^\ell = \infty] > 0, \text{ and } P_0\text{-a.s., } K < \infty, \text{ provided}$$
$$K \overset{\text{def}}{=} \inf\{k \geq 1, \, S_k < \infty \text{ and } R_k = \infty\}.$$

We can then define:

(4.25)
$$\tau_1 = S_K.$$

We can see (cf. [68]) that under $P_0$, $X_{\tau_1+\cdot} - X_{\tau_1}$ has same distribution as $X$. under $P_0[\cdot \,|\, D^\ell = \infty]$, and we can therefore iterate the definition

$$\tau_2 = \tau_1(X) + \tau_1(X_{\tau_1+\cdot} - X_{\tau_1}) \quad (= \infty \text{ if } \tau_1 = \infty)$$

and by induction

$$\tau_{k+1} = \tau_1(X) + \tau_k(X_{\tau_1+\cdot} - X_{\tau_1}),$$

and show the following renewal property:

(4.26)
under $P_0$, $(X_{\tau_1\wedge\cdot})$, $(X_{(\tau_1+\cdot)\wedge\tau_2} - X_{\tau_1}), \ldots, (X_{(\tau_k+\cdot)\wedge\tau_{k+1}} - X_{\tau_k}), \ldots$
are independent and except for the first variable distributed like
$(X_{\tau_1\wedge\cdot})$ under $P_0[\cdot \,|\, D^\ell = \infty]$.

Note incidentally that $\tau_{k+1} - \tau_k$ is indistinguishable from a measurable function of $(X_{(\tau_k+\cdot)\wedge\tau_{k+1}} - X_{\tau_k})$, (namely the first time when the trajectory remains from then on constant). One knows more than (4.26), we refer to [68] for more details. When $d = 1$, the above renewal structure can be found in Kesten [30] and also implicitly in Kesten-Kozlov-Spitzer [32]. We can now state the

**Condition (T):** *(relative to $\ell \in S^{d-1}$, $a > 0$),*

(4.27)
    i)   $P_0[\lim X_n \cdot \ell = \infty] = 1$

    ii)  *for some $c > 0$, $E_0[\exp\{c X^*_{\tau_1}\}] < \infty$, with $X^*_n = \sup_{0\leq k\leq n} |X_k|$.*

When $d = 1$, one can see, cf. Proposition 2.6 of [67], that ii) follows from i). Condition (T) thus characterizes transient random walks in random environment. In fact (cf. Solomon [61]), when $d = 1$, Condition (T) with respect to $\ell$ and $a$ is equivalent to $\mathbb{E}[\log \rho] \cdot \ell < 0$, in the notations of (0.1).

# Lecture 5:

# More on Random Walks in Random Environment

We further investigate model a) of the introduction. In particular, we discuss the important role of pockets of small eigenvalues in the slowdowns of random walks in random environment.

The coming proposition provides a wealth of examples of multi-dimensional random walks in random environment which fulfill condition (T). It shows that one has instances where (T) holds, in the three categories: non-nestling, marginal nestling, plain nestling, see (4.9), (4.10), (4.11).

**Proposition 5.1.** *Assume that for $\ell \in S^{d-1}$,*

(5.1)
$$\mathbb{E}[(d(0,\omega) \cdot \ell)_+] > \frac{1}{\kappa} \, \mathbb{E}[(d(0,\omega) \cdot \ell)_-],$$
*($\kappa$ is defined at the beginning of Lecture 1)*

*then* (T) *holds relative to $\ell$ and arbitrary $a > 0$.*

*Sketch of proof:* To prove (4.27) i) and ii), one needs good controls on the exit measures of $X_n$ from thick slabs under $P_0$. Although the behavior of $(X_n)$ under the annealed measure is in a sense simpler than under the quenched measure, one nevertheless looses the Markovian character of $(X_n)$ under the annealed measure. The lemma below, due to Kalikow [29], enables to come back to a Markovian setting. One first introduces certain auxiliary Markov chains. For $U \Subset \mathbb{Z}^d$, a connected subset containing 0, one defines an auxiliary Markov chain, with transition kernel

(5.2)
$$\widehat{P}_U(x, x+e) = \frac{\mathbb{E}[g_U(0, x, \omega)\omega(x, e)]}{\mathbb{E}[g_U(0, x, \omega)]}, \quad x \in U, \ |e| = 1,$$
$$\text{with } g_U(x, y, \omega) = E_{x,\omega}\Big[\sum_0^{T_U} 1\{X_n = y\}\Big]$$

$$\widehat{P}_U(x, x) = 1, \text{ when } x \in \partial U = \{z \in U^c, \exists z' \in U, |z' - z| = 1\}.$$

The state space of the chain is $U \cup \partial U$ and one denotes by $\widehat{P}_{x,U}$ its canonical law starting from $x$.

**Lemma 5.2.**

(5.3) *If $\widehat{P}_{0,U}[T_U < \infty] = 1$, then $P_0[T_U < \infty] = 1$ and $X_{T_U}$ has same law under $\widehat{P}_{0,U}$ and $P_0$.*

*Proof.* For $x \in U \cup \partial U$, $\omega \in \Omega$, an application of the simple Markov property shows that:

$$g_U(0, x, \omega) = \delta_{0,x} + \sum_{y \in U} g_U(0, y, \omega)\, \omega(y, x - y), \quad (\text{with } \omega(x, z) = 0, \text{ when } |z| \neq 1).$$

Thus taking $\mathbb{P}$-expectations:

$$\mathbb{E}[g_U(0, x, \omega)] = \delta_{0,x} + \sum_{y \in U} \mathbb{E}[g_U(0, y, \omega)\, \omega(y, x - y)]$$

(5.4)

$$\overset{(5.2)}{=} \delta_{0,x} + \sum_{y \in U} \mathbb{E}[g_U(0, y, \omega)]\, \widehat{P}_U(y, x - y).$$

Moreover:

(5.5)
$$\widehat{g}_U(x) = \widehat{E}_{0,U}\Big[\sum_{k=0}^{T_U} 1\{X_k = x\}\Big], \quad x \in U \cup \partial U,$$

is the minimal non-negative solution of the equation:

(5.6)
$$f(x) = \delta_{0,x} + \sum_{y \in U} f(y)\, \widehat{P}_U(y, x - y), \quad f : U \cup \partial U \to \mathbb{R}_+.$$

(Indeed, $\widehat{g}_{U,n}(x) \overset{\text{def}}{=} \widehat{E}_{0,U}\left[\sum_{k=0}^{T_U \wedge n} 1\{X_k = x\}\right]$ satisfies

$$\widehat{g}_{U,n+1}(x) = \delta_{0,x} + \sum_{y \in U} \widehat{g}_{U,n}(y)\, \widehat{P}_U(y, x - y),$$

and one easily sees by induction that $f \geq \widehat{g}_{U,n}$, so that letting $n$ tend to infinity, $f \geq \widehat{g}_U$). As a result:

(5.7)
$$\widehat{g}_U(x) \leq \mathbb{E}[g_U(0, x, \omega)], \quad x \in U \cup \partial U,$$

moreover on $\partial U$:

$$\widehat{g}_U(x) = \widehat{P}_{0,U}[T_U < \infty, X_{T_U} = x], \quad \mathbb{E}[g(0, x, \omega)] = P_0[T_U < \infty, X_{T_U} = x].$$

By assumption $\sum_{\partial U} \widehat{g}_U(x) = 1$, and from (5.7)

$$\widehat{g}_U(x) = \mathbb{E}[g_U(0, x, \omega)], \quad x \in \partial U.$$

The claim now follows.                                             $\square$

Further for $U \Subset \mathbb{Z}^d$, connected and containing 0, one defines the auxiliary drift:

(5.8)
$$\widehat{d}_U(x) = \widehat{E}_{x,U}[X_1 - X_0], \quad x \in U \cup \partial U.$$

Now for the $\ell \in S^{d-1}$ which appears in (5.1),

$$\widehat{d}_U(x) \cdot \ell \overset{(5.2)}{=} \frac{\mathbb{E}[g_U(0, x, \omega)\, d(x, \omega) \cdot \ell]}{\mathbb{E}[g_U(0, x, \omega)]},$$

and by a classical Markov chain calculation, for $x \in U$,

$$(5.9) \qquad g_U(0, x, \omega) = \frac{P_{0,\omega}[H_x < T_U]}{P_{x,\omega}[\tilde{H}_x > T_U]} = \frac{P_{0,\omega}[H_x < T_U]}{\sum\limits_{|e|=1} \omega(x, e) \, P_{x+e,\omega}[H_x > T_U]}$$

where $\tilde{H}_x = \inf\{n \geq 1, X_n = x\}$ is the hitting time of $\{x\}$. As a result:

$$(5.10)$$

$$\hat{d}_U(x) \cdot \ell \overset{(5.9)}{=} \frac{1}{\mathbb{E}[g_U(0, x, \omega)]} \; \mathbb{E}\left[ \frac{P_{0,\omega}[H_x < T_U]}{\sum\limits_{|e|=1} \omega(x, e) \, P_{x+e,\omega}[H_x > T_U]} \, d(x, \omega) \cdot \ell \right]$$

$$\geq \frac{1}{\mathbb{E}[g_U(0, x, \omega)]} \; \mathbb{E}\left[ \frac{P_{0,\omega}[H_x < T_U]}{\max\limits_{|e|=1} P_{x+e,\omega}[H_x > T_U]} \left( (d(x, \omega) \cdot \ell)_+ - \frac{1}{\kappa} (d(x, \omega) \cdot \ell)_- \right) \right],$$

note that the first ratio under the $\mathbb{P}$-expectation is independent of $\omega(x, \cdot)$, so the above quantity equals

$$= \; \mathbb{E}[(d(x, \omega) \cdot \ell)_+ - \frac{1}{\kappa} (d(x, \omega) \cdot \ell)_-)] \frac{1}{\mathbb{E}[g_U(0, x, \omega)]} \; \mathbb{E}\left[ \frac{P_{0,\omega}[H_x < T_U]}{\max\limits_{|e|=1} P_{x+e,\omega}[H_x > T_U]} \right]$$

$$\overset{(5.9)}{\geq} \; \kappa \, \mathbb{E}\left[ \left( (d(0, \omega) \cdot \ell)_+ - \frac{1}{\kappa} (d(0, \omega) \cdot \ell)_- \right) \right].$$

In other words, we see that

$$(5.11) \qquad \epsilon(\ell, \mu) = \inf_{U, x \in U} \hat{d}_U(x) \cdot \ell > 0 \quad (U \subsetneq \mathbb{Z}^d, \text{ connected}, 0 \in U).$$

This is **Kalikow's condition**.

It is then not too hard to see (cf. Lemma 1.1 of [66]), that for $\theta(\epsilon) > 0$, and $U$ arbitrary as above:

$$(5.12) \qquad \exp\{-\theta X_n \cdot \ell\} \text{ is a } \hat{P}_{x,U}\text{-supermartingale}, \; x \in U \cup \partial U.$$

Then as follows with the help of (5.3) and (5.12):

$$(5.13) \qquad \lim_{L \to \infty} P_0[T_{U_{+,L}} < \infty, \, X_{T_{U_{+,L}}} \cdot \ell > 0] > 0, \; U_{+,L} \overset{\text{def}}{=} \{0 \leq x \cdot \ell < L\}$$

$$(5.14) \qquad \lim_{L \to \infty} P_0[T_{U_L} < \infty, \, X_{T_{U_L}} \cdot \ell > 0] = 1, \; U_L \overset{\text{def}}{=} \{|x \cdot \ell| < L\}.$$

From (5.13) and (5.14) we respectively see that $P_0[D^\ell = \infty] > 0$ and $P_0[\overline{\lim} X_n \cdot \ell = \infty] = 1$. From the $0 - 1$ law, see (4.21), we also known that $P_0[\{\lim X_n \cdot \ell = \infty\} \cup \{\lim X_n \cdot \ell = -\infty\}] = 0$ or $1$. Since $P_0[D^\ell = \infty] > 0$, this probability equals 1, and since $P_0[\overline{\lim} X_n \cdot \ell = \infty] = 1$, as a matter of fact:

$$(5.15) \qquad P_0[\lim X_n \cdot \ell = \infty] = 1, \; (\text{i.e. i) of condition (T)}).$$

We briefly sketch how (T) ii) is proved. We only discuss the estimate

(5.16) $$E_0[\exp\{c\,X_{\tau_1}\cdot\ell\}] < \infty, \text{ for } c > 0 \text{ small},$$

(for more details see [66]). The above quantity equals

$$\sum_{k\geq 1} E_0[\exp\{c\,X_{S_k}\cdot\ell\}, \ S_k < \infty, \ D^\ell\circ\theta_{S_k} = \infty]$$

using the fact that $X_{S_k}\cdot\ell \leq a + 1 + M_{k-1}$ and a renewal argument

$$\leq \text{const} \sum_{k\geq 1} E_0[\exp\{c(a+1+\overline{M})\}], \ D^\ell < \infty]^{k-1},$$

where

(5.17) $$\overline{M} = \sup\{X_n\cdot\ell - X_0\cdot\ell, \ 0 \leq n \leq D^\ell\}.$$

The crucial estimate is then:

(5.18) $$E_0[\exp\{c\,\overline{M}\}, \ D^\ell < \infty] < 1, \text{ for small } c > 0.$$

The quantity in the left hand side of (5.18) is smaller than

$$\sum_{m\geq 0} e^{c2^{m+1}} P_0[2^m \leq \overline{M} < 2^{m+1}, \ D^\ell < \infty] + e^c P_0[0 \leq \overline{M} \leq 1, \ D^\ell < \infty]$$

and since $P_0[D^\ell < \infty] < 1$, (5.18) will follow from:

(5.19) $$P_0[2^m \leq \overline{M} < 2^{m+1}] \leq \exp\{-\text{const } 2^m\}, \text{ for large } m.$$

To prove this last point, one uses the auxiliary martingales (under $\widehat{P}_{0,U}$):

$$M_n^U = X_n - X_0 - \sum_0^{n-1} \widehat{d}_U(X_k),$$

as well as the inclusion:

(5.20) $$\{2^m \leq \overline{M} < 2^{m+1}\} \subseteq \{T_{2^m}^\ell < \infty\} \cap \{\widetilde{T}_0^\ell \circ \theta_{T_{2^m}^\ell} < T_{2^{m+1}}^\ell \circ \theta_{T_{2^m}^\ell}\}.$$

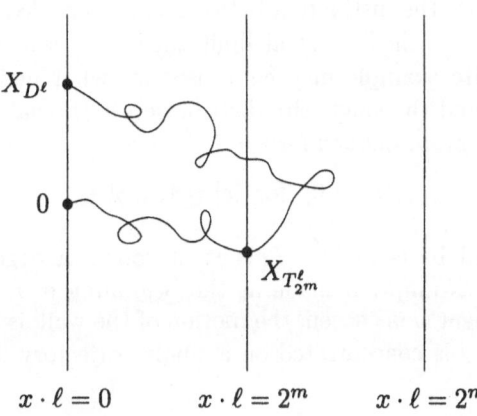

$$x\cdot\ell = 0 \qquad x\cdot\ell = 2^m \qquad x\cdot\ell = 2^{m+1}$$

With the help of Kalikow's condition (5.11), and Azuma's inequality (1.22), one sees that $\widehat{P}_{0,U}[T^{\ell}_{2^m} > \text{const } 2^m] \le e^{-c' 2^m}$, for $U = \{x \cdot \ell < 2^m\}$. Then with (5.3) one finds $P_0[\,|X_{T^{\ell}_{2^m}}| \ge \text{const } 2^m] = \widehat{P}_{0,U}[\,|X_{T_U}| \ge \text{const } 2^m] \le e^{-c' 2^m}$, and (5.19) follows rather easily from (5.20) and the strong Markov property, together with the fact that $P_0[\widetilde{T}_{-2^m} < T^{\ell}_{2^m}] \le e^{-\theta 2^m}$, (cf. (5.3), (5.12)). □

When $d = 1$, (T) may hold in situations where the asymptotic velocity vanishes (for instance when $\mathbb{E}[\log \rho] < 0$, but $\mathbb{E}[\rho] \ge 1$). It may also happen that the asymptotic velocity does not vanish but $X_n$ does not fulfill a central limit theorem, see Kesten-Kozlov-Spitzer [32]. The situation is quite different when $d \ge 2$, cf. [67]:

**Theorem 5.3.** *($d \ge 2$, under Condition (T) )*

(5.21) $P_0$-*a.s.* $\dfrac{X_n}{n} \to v$ *($\neq 0$, deterministic)*

(5.22) $(B^n_t) = \dfrac{X_{[nt]} - [nt]\, v}{\sqrt{n}}$ *converges in law (on $D(\mathbb{R}_+, \mathbb{R}^d)$), to a Brownian motion with non-degenerate covariance matrix $A$.*

*Sketch of Proof:* With the help of the renewal property (cf. (4.26)), the law of large numbers in essence follows from

$$E_0[\tau_1 \,|\, D^\ell = \infty] < \infty\,, \tag{5.23}$$

(and in fact $v = \frac{E_0[X_{\tau_1} \,|\, D^\ell = \infty]}{E_0[\tau_1 \,|\, D^\ell = \infty]}$). In the same vein, to prove (5.22), it suffices to show that

$$E_0[\tau_1^2 \,|\, D^\ell = \infty] < \infty\,, \tag{5.24}$$

(and in fact $A = \frac{E_0[(X_{\tau_1} - \tau_1 v)\, {}^t(X_{\tau_1} - \tau_1 v) \,|\, D^\ell = \infty]}{E_0[\tau_1 \,|\, D^\ell = \infty]}$). We discuss further below how (5.23) and (5.24) are obtained.

**Remark:**

**1)** The reader should ponder the distinction between the "annealed" central limit theorem (5.22) and the "quenched" central limit theorem discussed in Lecture 2. The following degenerate example may help visualize what may be at stake. Consider the case $d = 2$, and the single site distribution is the half-sum of Dirac masses on the two vectors $p_1, p_2$ defined for $i = 1, 2$, via:

$$p_i(e_i) = 1, \;\; p_i(e) = 0, \;\; \text{for } |e| = 1,\; e \neq e_i\,,$$

with $(e_1, e_2)$ the canonical basis of $\mathbb{R}^2$. This is of course a degenerate situation where the ellipticity assumption made at the beginning of Lecture 1 is violated. Once the environment $\omega$ is chosen, the motion of the walk is deterministic; the quenched measure $P_{0,\omega}$ is concentrated on a single trajectory dictated by $\omega$.

It is easy to see that (5.21) holds with $v = \frac{1}{2}(e_1 + e_2)$. Further the annealed central limit theorem (5.22) holds (the limiting Brownian motion has a degenerate covariance matrix corresponding to fluctuations purely in the $e_1 - e_2$ direction). On the other hand $B^n_\cdot$ does not satisfy a central limit theorem under $P_{0,\omega}$ for $\mathbb{P}$-typical $\omega$.

**2)** Incidentally one can show, cf. Theorem 2.2. of [67], that when (T) holds relative to $\ell$ and $a$, there exists a deterministic direction $\hat{v}$ (none other than $\frac{v}{|v|}$, when $d \geq 2$), such that (T) holds relative to $\ell'$, $a'(> 0)$ if and only if: $\ell' \cdot \hat{v} > 0$.

**3)** If one assumes Kalikow's condition (5.11), there is a quick argument to prove (5.23); in essence the ballistic nature of the walk can rather straightforwardly be extracted from (5.11), cf. Lemma 2.2 of Sznitman-Zerner [68].                          □

As a matter of fact one has much stronger estimates than (5.23), and (5.24), cf. [67]:

**Theorem 5.4.** *($d \geq 2$, under Condition* (T) *)*

$$(5.25) \qquad P_0[\tau_1 > u] \leq \exp\{-(\log u)^\alpha\}, \text{ for large } u, \text{ when } \alpha < \frac{2d}{d+1}$$

*(thus all moments of $\tau_1$ are finite).*

Note that when $d = 1$, even when (5.23) holds, (5.24) may fail.

In a sense $u \to P_0[\tau_1 > u]$ plays an analogous role to the annealed survival probability of the simple random walk among random traps: $n \to S(n)$, cf. (3.4). In particular one way for $\tau_1$ to be large is that there is a trap in the neighborhood of 0, where the particle spends much time and then visits $\{x \cdot \ell \leq 0\}$. With this remark one thus sees in the plain nestling situation that in the notations of (4.13):

$$(5.26) \qquad \begin{aligned} &\varliminf_{u \to \infty} (\log u)^{-d} \log P_0[\tau_1 > u] \geq \\ &\varliminf_{n \to \infty} (\log n)^{-d} \log \mathbb{E}[\mathcal{T}_L, P_{0,\omega}[T_{B_L} > n, X_n \cdot \ell \leq 0]] > -\infty, \end{aligned}$$

provided $L = c \log n$, with $c$ large. It is an open problem to determine whether more precisely than (5.25), $\varlimsup (\log u)^{-d} \log P_0[\tau_1 > u] < 0$.

We shall not prove (5.25) here. We shall rather explain why a certain type of large deviation estimates on the exit distribution of the particle from large slab plays an analogous role to the key control (3.22) on the occurrence of small eigenvalues, when one instead investigates the asymptotic behavior of $S(n)$. We need some notations. For $\beta \in (0, 1]$, $L > 0$, we define

$$U_{\beta, L} = \{x \in \mathbb{Z}^d, x \cdot \ell \in (-L^\beta, L)\}, \text{ so } U_{1,L} = U_L \text{ in the notations of (5.14)}.$$

We are interested in the $\mathbb{P}$-probability that $P_{0,\omega}[X_{T_{U_{\beta,L}}} \cdot \ell > 0]$ is "atypically small".

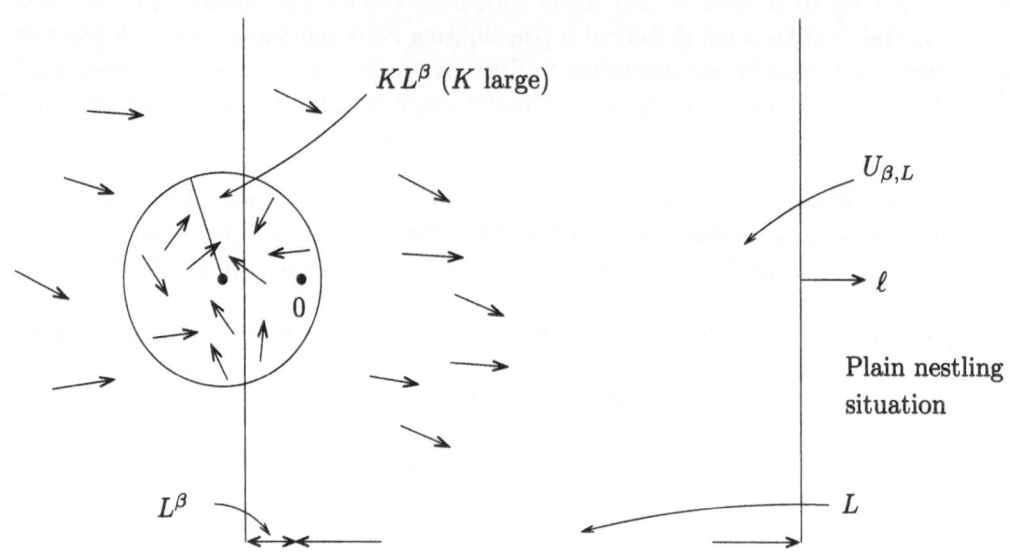

$KL^{\beta}$ ($K$ large)

$U_{\beta,L}$

$\ell$

Plain nestling situation

$L^{\beta}$

$L$

"An event under which $P_{0,\omega}[X_{T_{U_{\beta,L}}} \cdot \ell > 0] \leq e^{-cL^{\beta}}$".

**Proposition 5.5.** *($d \geq 2$ and (T) holds relative to $\ell$ and $a$).*

*Assume $\beta \in (0,1)$ is such that for all $c > 0$,*

$$(5.27) \qquad \varlimsup_{L \to \infty} \frac{1}{L} \log \mathbb{P}[P_{0,\omega}[X_{T_{U_{\beta,L}}} \cdot \ell > 0] \leq e^{-cL^{\beta}}] < 0, \quad then$$

$$(5.28) \qquad \varlimsup_{u \to \infty} (\log u)^{-\frac{1}{\beta}} \log P_0[\tau_1 > u] < 0.$$

*Proof.* We define

$$(5.29) \qquad \Delta(u) = \delta(\log u), \quad and \quad L(u) = \Delta(u)^{\frac{1}{\beta}} \, (\gg \Delta(u)),$$

where $\delta > 0$ is small and defined below. Then for large $u$

$$(5.30) \qquad P_0[\tau_1 > u] \leq P_0[\tau_1 > u, \, T_{C_{L(u)}} \leq \tau_1] + P_0[T_{C_{L(u)}} > u],$$

where $C_r = (-\frac{r}{2}, \frac{r}{2})^d$. Observing that $T_{C_{L(u)}} \leq \tau_1$ forces $\sup_{k \leq \tau_1} |X_k| \geq \frac{L(u)}{2}$, the application of (T) ii) and Chebyshev's inequality shows that

$$\leq \exp\{-\text{const } L(u)\} + P_0[T_{C_{L(u)}} > u].$$

We thus only need to concentrate on the last term. For $U \subseteq \mathbb{Z}^d$ non-empty, $\omega \in \Omega$, we introduce

$$t_\omega(U) = \inf \left\{ n \geq 0, \ \sup_U P_{x,\omega}[T_U > n] \leq \tfrac{1}{2} \right\} \leq \infty,$$

(note that $t_\omega(U)$ is always finite for finite $U$), then

$$P_0[T_{C_{L(u)}} > u] \leq \mathbb{E}\left[P_{0,\omega}[T_{C_{L(u)}} > u], \ t_\omega(C_{L(u)}) \leq \frac{u}{(\log u)^{\frac{1}{\beta}}}\right]$$

$$+ \ \mathbb{P}\left[t_\omega(C_{L(u)}) > \frac{u}{(\log u)^{\frac{1}{\beta}}}\right] \overset{\substack{\text{Markov} \\ \text{property}}}{\leq} \left(\frac{1}{2}\right)^{[(\log u)^{\frac{1}{\beta}}]} + \mathbb{P}\left[t_\omega(C_{L(u)}) > \frac{u}{(\log u)^{\frac{1}{\beta}}}\right].$$

Observe that $\sup_{x \in C_{L(u)}} P_{x,\omega}[T_{C_{L(u)}} \geq t_\omega(C_{L(u)})] \geq \tfrac{1}{2}$, and therefore when $t_\omega(C_{L(u)}) > \frac{u}{(\log u)^{\frac{1}{\beta}}}$, necessarily for some $x_1 \in C_{L(u)}$:

$$\frac{1}{2} \frac{u}{(\log u)^{\frac{1}{\beta}}} \leq E_{x_1,\omega}[T_{C_{L(u)}}] = \sum_{y \in C_{L(u)}} \frac{P_{x,\omega}[H_y < T_{C_{L(u)}}]}{P_{y,\omega}[\widetilde{H}_y > T_{C_{L(u)}}]}.$$

by a similar calculation as in (5.9). So for some $x_2 \in C_{L(u)}$:

$$P_{x_2,\omega}[\widetilde{H}_{x_2} > T_{C_{L(u)}}] \leq \frac{|C_{L(u)}|}{2u} (\log u)^{\frac{1}{\beta}} \quad \left(\sim \frac{1}{u} \text{ up to logarithmic corrections}\right)$$

Now for arbitrary $x \neq x_2$:

$$P_{x_2,\omega}[\widetilde{H}_{x_2} > T_{C_{L(u)}}] \geq P_{x_2,\omega}[\widetilde{H}_{x_2} > H_x] \, P_{x,\omega}[H_{x_2} > T_{C_{L(u)}}],$$

and choosing $x \asymp x_2 + 2\Delta(u)\,\ell$, we see that provided $\delta = \delta(\kappa)$ is chosen small in (5.29), then for large $u$:

$$\frac{1}{\sqrt{u}} \geq P_{x,\omega}[H_{x_2} > T_{C_{L(u)}}] \geq P_{x,\omega}[X_{T_{x+U_{\beta,L(u)}}} \cdot \ell > x \cdot \ell]$$

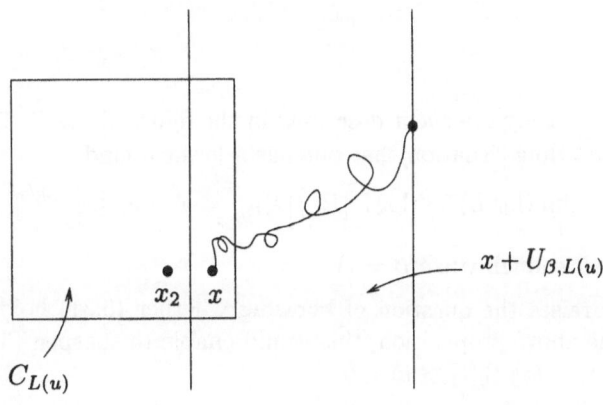

However $\frac{1}{\sqrt{u}} \le e^{-c'\Delta(u)} = e^{-c'\,L(u)^\beta}$, so that using (5.27) and translation invariance, for large $u$:

$$P_0[T_{C_{L(u)}} > u] \quad \le \quad \left(\frac{1}{2}\right)^{[(\log u)^\beta]} + |C_{L(u)}|\,\mathbb{P}[P_{0,\omega}[X_{T_{U_{\beta,L(u)}}} \cdot \ell > 0] \le e^{-c'\,L(u)^\beta}]$$

$$\overset{(5.27)}{\le} \quad \exp\{-c''\,L(u)\}.$$

coming back to (5.30), our claim follows.                                      $\square$


The crucial estimate is then (cf. Theorem 3.4 of [67]):

**Theorem 5.6.** *($d \ge 2$, under Condition (T) )*

*For $c > 0$, $\beta \in \left(\frac{1}{2}, 1\right]$*

(5.31)     $\overline{\lim}_L L^{-\alpha} \log \mathbb{P}[P_{0,\omega}[X_{T_{U_{\beta,L}}} \cdot \ell > 0] \le e^{-cL^\beta}] < 0, \quad \text{for } \alpha < d(2\beta - 1),$

*and one can choose $\alpha = d$, when $\beta = 1$.*

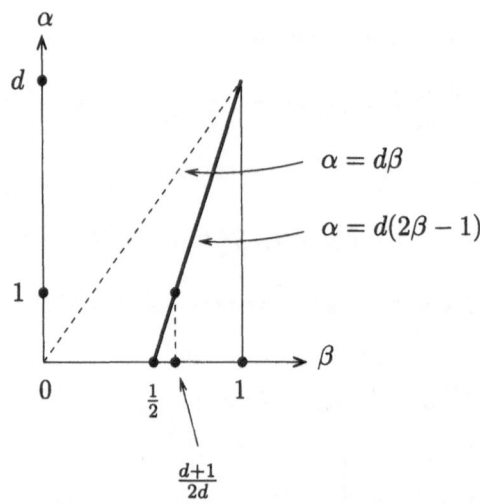

**Remarks:**

**1)** Note that using the event described in the figure above (5.27), one can show in the plain nestling situation that one has a lower bound

$$\lim_L (\log L)^{-d\beta} \log \mathbb{P}[P_{0,\omega}[X_{T_{U_{\beta,L}}} \cdot \ell > 0] \le e^{-cL^\beta}] > -\infty,$$

($c > 0$ small enough when $\beta = 1$).

This raises the question of knowing whether (5.31) holds with $\alpha = d\beta$. In view of the above proposition, this would enable to sharpen (5.25) and show that $\underline{\lim}_L (\log u)^{-d} \log P_0[\tau_1 > u] < 0$.

**2)** The walks which are "neutral or biased to the right", constitute an interesting sub-class of the marginal nestling case, corresponding to the situation where for some $\delta > 0$,

$$\mathbb{P}\left[\left\{\omega(0\cdot) \equiv \frac{1}{2d}\right\} \quad \cup \quad \{d(0,\omega) \cdot e_1 \geq \delta\}\right] = 1$$

$$\underset{\text{neutral case}}{\uparrow} \qquad\qquad \underset{\text{case biased to the right}}{\uparrow}$$

(5.32)

$$\mathbb{P}\left[\omega(0,\cdot) \equiv \frac{1}{2d}\right] > 0, \qquad \mathbb{P}[d(0,\omega) \cdot e_1 \geq \delta] > 0.$$

In this setting (5.1) and thus Kalikow's condition and Condition (T) hold. One can then show (cf. Theorem 2.5 of [66]):

**Theorem 5.7.** *(d $\geq$ 1, case neutral or biased to the right)*

(5.33) $\qquad -\infty < \underline{\lim_{u}} \, u^{-\frac{d}{d+2}} \, \log P_0[\tau_1 > u] \leq \overline{\lim_{u}} \, u^{-\frac{d}{d+2}} \, \log P_0[\tau_1 > u] < 0\,,$

*(this is stated in [66] for $\ell = e_1$, a > 0, but holds also for $\ell$ with $\ell \cdot v > 0$).*

This of course strongly reminds of (3.11). The role of the key estimate (5.31) is played by (cf. Proposition 3.1 of [65]).

**Theorem 5.8.** *(d $\geq$ 1, case neutral or biased to the right)*

*There exists $p_0 \in (\frac{1}{2}, 1)$ such that*

(5.34) $\qquad\qquad \overline{\lim_{L}} \, L^{-d} \, \log \mathbb{P}[P_{0,\omega}[X_{T_{U_L}} \cdot e_1 > 0] \leq p_0] < 0\,,$

*(here $U_L = \{|x \cdot e_1| < L\}$).*

**3)** In the non-nestling situation, when $K_0 \subset \{\ell \cdot x > 0\}$, we can see that $\tau_1$ has an exponential tail (cf. Theorem 2.1 of [66]).

The estimates (5.25) and (5.33) have several applications to the analysis of slowdowns of random walks in random environment. It turns out that in the nestling case (i.e. when $0 \in K_0$), the segment $[0, v] \subseteq \mathbb{R}^d$, plays a critical role, when $d \geq 2$ and (T) holds, for the annealed large deviation estimates on $\frac{X_n}{n}$.

**Theorem 5.9.** *([67], d $\geq$ 2, under (T) )*

*When $\mathcal{O}$ is an open neighborhood of $[0, v]$,*

(5.35) $\qquad\qquad\qquad \overline{\lim} \, \frac{1}{n} \, \log P_0\left[\frac{X_n}{n} \notin \mathcal{O}\right] < 0$

*(in the non-nestling case one can pick $\mathcal{O}$ to be a neighborhood of $v$).*

*Further, when $0 \in K_0$ (i.e. in the nestling case), $[0, v]$ is "critical":
for $\mathcal{U}$ open with $\mathcal{U} \cap [0, v] \neq \emptyset$,*

(5.36)
$$\varliminf_{n} \frac{1}{n} \log P_0\left[\frac{X_n}{n} \in \mathcal{U}\right] = 0.$$

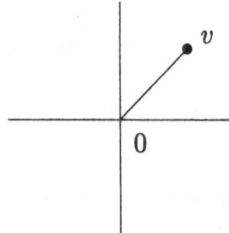

**Remarks:**

**1)** When $d = 1$, (with $v \neq 0$), this result goes back to Dembo-Peres-Zeitouni [15]. An annealed large deviation principle is even proved in Comets-Gantert-Zeitouni [12].

**2)** When $d \geq 1$, Zerner [73] has shown in the nestling case a large deviation principle under the quenched measure for $\mathbb{P}$-a.e. $\omega$. When $d \geq 2$, it is not known in general where the null-set of the deterministic rate function $I(\cdot)$ governing the large deviation principle is located. However with the help of the above result, one can see that when (T) holds, $\{I = 0\} = [0, v]$.

The study of slowdowns corresponds to the investigation of large deviations of $\frac{X_n}{n}$ in the neighborhood of the critical segment $[0, v]$ (under (T), in the nestling case). In particular, cf. Remark 4.5 of [67].

**Theorem 5.10.** *($d \geq 1$, case neutral or biased to the right, compare with (3.11) )*

- *For $\mathcal{U}$, as in (5.36),*

(5.37)
$$\varliminf_{n} n^{-\frac{d}{d+2}} \log P_0\left[\frac{X_n}{n} \in \mathcal{U}\right] > -\infty.$$

- *When $\mathcal{O}$ is a neighborhood of $v$*

(5.38)
$$\varlimsup_{n} n^{-\frac{d}{d+2}} \log P_0\left[\frac{X_n}{n} \notin \mathcal{O}\right] < 0.$$

*Moreover when $\delta \in (0, 1)$,*

(5.39)
$$\log P_0\left[\left|\frac{X_n}{n}\right| < \delta v\right] \asymp \log \mathbb{E}[\exp\{-n\,\lambda_\omega([-n, n]^d)\}],$$

*where "$\asymp$" means that the ratio of the two quantities remains bounded for large $n$.*

The left hand side of (5.39) measures slowdowns and the right hand side the influence of small principal Dirichlet eigenvalues.

When $d = 1$, (5.37), (5.38) was first proved in Dembo-Peres-Zeitouni [15]. An annealed large deviation principle, on the critical segment $[0, v]$, $(d = 1)$, was then proved in Pisztora-Povel-Zeitouni [50]. The quenched case, $(d = 1)$, was later treated in Pisztora-Povel [49]. When $d \geq 1$, the quenched case roughly behaves like the quenched asymptotics of the long time survival of a particle diffusing among soft obstacles, see for instance the end of Section V in [66] as well as Chapter 4 §5 of [64].

In the plain nestling situation it is less costly for the walk to produce slowdowns, thanks to the presence of stronger traps, and one has (cf. Theorem 4.3 of [67]).

**Theorem 5.11.** *($d \geq 2$, under* (T) *)*

- *In the plain nestling case, for $\mathcal{U}$ as in (5.36)*

$$(5.40) \qquad \varliminf_{n} \, (\log n)^{-d} \log P_0\left[\frac{X_n}{n} \in \mathcal{U}\right] > -\infty \, .$$

- *In general, for $\mathcal{O}$ as in (5.38)*

$$(5.41) \qquad \varlimsup_{n} \, (\log n)^{-\alpha} \log P_0\left[\frac{X_n}{n} \notin \mathcal{O}\right] < 0, \ \textit{for} \ \alpha < \frac{2d}{d+1} \, .$$

**Remarks:**

**1)** When $d = 1$, the plain nestling case for $v \neq 0$ is treated in Dembo-Peres-Zeitouni [15]. Roughly speaking the probabilities of slowdown decay like a power of $n$.

**2)** It is an open problem whether (5.41) holds with $\alpha = d$. This is closely related to the question of knowing whether (5.31) holds for $\alpha = d\beta$.

**3)** One has analogous upper and lower bounds for the quantity $\mathbb{E}[\exp\{-n\,\lambda_\omega([-n,n]^d)\}]$, which measures the influence of "small eigenvalues". $\square$

# PART TWO

by Erwin Bolthausen, Universität Zürich

## LECTURES ON SPIN GLASSES

# Lecture 6:

## On the Sherrington–Kirkpatrick Model of Spin Glasses

We fix some notational conventions. If $(a_N)$ and $(b_N)$ are two sequences of positive real numbers, we write $a_N \asymp b_N$ if for all $\varepsilon > 0$

$$e^{-\varepsilon N} b_N \leq a_N \leq e^{\varepsilon N} b_N$$

for all large enough $N$ ($N \geq N_o(\varepsilon)$). We write $a_N \ll b_N$ if there exists $\varepsilon > 0$ such that

$$a_N \leq e^{-\varepsilon N} b_N$$

for large enough $N$. $a_N \sim b_N$ has the usual meaning:

$$\lim_{N \to \infty} \frac{a_N}{b_N} = 1.$$

If $A_N$ and $B_N$ are sequences of random variables, taking values in the positive real numbers, then $A_N \sim B_N$ or $A_N \asymp B_N$ just means that the above relations hold almost surely.

$N$ will always be a parameter measuring the "size" of our system, and will in the end tend to $\infty$. Generally, all inequalities we write down are required to hold only for large enough $N$, without special mentioning, where what "large enough" means may depend on other parameters involved (typically $\beta$, $h$ etc.)

We also fix some convention about constants. $C$ will be a generic positive constant not necessarily the same at different occurrences. If it depends on other parameters, like $\beta$, then we will indicate that clearly be writing for instance $C(\beta)$.

Probability measures will be denoted by $P$ or $\mathbb{P}$. The latter is used for probability measures governing the random environment. $E$, $\mathbb{E}$ are used for the expectations. If $X$ is a random variable and $A$ an event, we sometimes write $E(X; A)$ instead of $\int_A X \, dP$.

If $M$ is a finite set, then $\#M$ denotes the number of elements in $M$.

### 6.1. Gibbs Measures with Pair Interactions

Let $\Lambda$ be a finite set, and $\Sigma_\Lambda = \{-1, 1\}^\Lambda$. We will consider special probability measures on this set given by pair interactions. Let $J = \{J_{i,j}\}_{i,j \in \Lambda, i \neq j}$ be any collection of real numbers, having the property $J_{i,j} = J_{j,i}$, and let $h = \{h_i\}_{i \in \Lambda}$ be another collection of numbers.

We define the Gibbs measure (with free boundary conditions) to be the probability measure $P$ on $\Sigma_\Lambda$ by

$$(6.1) \qquad P_{J,\beta,h,\Lambda}(\sigma) = \frac{1}{Z_{J,\beta,h,\Lambda}} 2^{-\#\Lambda} \exp\left[\frac{\beta}{2} \sum_{\substack{i,j \in \Lambda \\ i \neq j}} J_{i,j}\sigma_i\sigma_j + \sum_{i \in \Lambda} h_i\sigma_i\right],$$

where $\beta > 0$ is the so-called inverse temperature. Usually, in physics literature, the $\beta$ factor is also applied to the second summand in the exponent $\sum_{i \in \Lambda} h_i\sigma_i$, but it is more convenient here not to do so. $Z_{J,\beta,h,\Lambda}$ is the normalization constant in order to make this a probability measure:

$$Z_{J,\beta,h,\Lambda} = \sum_{\sigma \in \Sigma_\Lambda} 2^{-\#\Lambda} \exp\left[\frac{\beta}{2} \sum_{\substack{i,j \in \Lambda \\ i \neq j}} J_{i,j}\sigma_i\sigma_j + \sum_{i \in \Lambda} h_i\sigma_i\right]$$

$Z$ is usually called the *partition function*. We will often drop some of the indices if there is no danger of confusion. The factor $2^{-\#\Lambda}$ is of course completely superfluous in the definition of the Gibbs measure, as it cancels out. It is however convenient to write $Z$ as an expectation under the coin tossing measure, which we usually will denote by $P_o$.

Without the $J$-interaction, $P$ is just coin tossing with biased coins. With the $J$-variables, the measure becomes much more interesting. It evidently gives higher weight to configurations $\sigma$ which have the property that $\sigma_i$ and $\sigma_j$ have the same sign if $J_{i,j} > 0$ and vice versa if it is negative. It should however be remarked that there might be conflicts. Take for instance $\Lambda$ consisting of three points, say $a, b, c$, and assume that $J_{a,b} = J_{a,c} = 1$, and $J_{b,c} = -1$. According the first two settings, $\sigma_b$ and $\sigma_c$ would like to align with $\sigma_a$, but according to the choice of $J_{a,c}$, $\sigma_b$ and $\sigma_c$ would prefer to have opposite signs. When such situations occur, one calls the system frustrated.

Typically, one is interested in large systems, so one considers sequences of sets $\Lambda_N$, where $\#\Lambda_N \to \infty$ as $N \to \infty$. This limit is called the "thermodynamic limit". A quantity of crucial importance is the so called *free energy*:

$$f(\beta, h) \stackrel{\text{def}}{=} \lim_{N \to \infty} \frac{1}{\#\Lambda_N} \log Z_{\beta,h,\Lambda_N},$$

if this limit exists. Unfortunately, there are very simple cases, where there is no proof that this limit exists. For instance, this is unknown on a level of mathematical rigor in the Sherrington-Kirkpatrick model at low temperature.

Most quantities of interest in the thermodynamic limit can be derived from the free energy. As an example, consider the so called mean magnetization $E(\frac{1}{\#\Lambda} \sum_{i \in \Lambda} \sigma_i)$. For $s \in \mathbb{R}$ consider a perturbed external field $h + s \stackrel{\text{def}}{=} \{h_i + s\}_{i \in \Lambda}$. Then evidently

$$E\left(\frac{1}{\#\Lambda} \sum_{i \in \Lambda} \sigma_i\right) = \frac{1}{\#\Lambda} \frac{d}{ds} \log E_o\left(\exp\left[\frac{\beta}{2} \sum_{i,j \in \Lambda} J_{i,j}\sigma_i\sigma_j + \sum_{i \in \Lambda}(h_i + s)\sigma_i\right]\right)\Big|_{s=0}.$$

If we are allowed to interchange limits (which of course needs justification), we get

$$\lim_{N \to \infty} E\left(\frac{1}{\#\Lambda_N} \sum_{i \in \Lambda_N} \sigma_i\right) = \frac{d}{ds} f(\beta, h+s)\big|_{s=0}.$$

### 6.2. The Curie–Weiss Model

The Curie-Weiss model is obtained by taking $J_{i,j} \equiv \frac{1}{N}$, $i \neq j$, and for simplicity $h_i \equiv 0$. It is the simplest of all models which exhibit a phase transition . A possible structure of $\Lambda$ is evidently playing no role. So we can as well assume that $\Lambda = \{1, \ldots, N\}$. In this case

$$P_{\beta,N}(\sigma) = \frac{1}{Z_{\beta,N}} 2^{-N} \exp\left[\beta \frac{N}{2}\left(\frac{1}{N}\sum_{i=1}^{N}\sigma_i\right)^2\right].$$

We evaluate the free energy and the spontaneous magnetization.
Let $S_N \stackrel{\text{def}}{=} \frac{1}{N}\sum_{i=1}^{N}\sigma_i$.

**Theorem 6.1.** *The free energy of the Curie-Weiss model is given by*

   a) *If $\beta \leq 1$ then $f(\beta) = 0$.*
   b) *If $\beta > 1$ then $f(\beta) = \log\cosh(\beta\xi) - \frac{\beta\xi^2}{2}$, where $\xi = \xi(\beta) > 0$ is the unique positive solution of the equation*

(6.2) $$\xi = \tanh(\beta\xi).$$

There are two instructive proofs for that. The first is based on large deviations for basic coin tossing, which can easily be extracted from Stirlings formula:

**Exercise 6.2.** *We denote by $P_o$ the standard coin tossing law for the $\sigma_i$. Then*

$$P_o(S_N = x) \asymp \exp[-NI(x)],$$

*uniformly in $x \in \{-1, \frac{-N+2}{N}, \frac{-N+4}{N}, \ldots, 1\}$, where*
$I(x) = \frac{1+x}{2}\log(1+x) + \frac{1-x}{2}\log(1-x)$.

From this exercise, we have

$$Z_{\beta,N} = E_o \exp\frac{\beta N}{2}S_N^2 =$$
$$= \sum_x e^{\beta N x^2/2} P_o(S_N = x) \asymp \sum_x \exp N\left[\frac{\beta x^2}{2} - I(x)\right].$$

In this sum, for a rough asymptotics, only the contribution of the sum counts which is coming from $x$ where $\frac{\beta x^2}{2} - I(x)$ is maximal:

$$\exp\left[N\sup_x\left(\frac{\beta x^2}{2} - I(x)\right)\right] \leq \sum_x \exp\left[N\left(\frac{\beta x^2}{2} - I(x)\right)\right]$$
$$\leq N\exp\left[N\sup_x\left(\frac{\beta x^2}{2} - I(x)\right)\right],$$

i.e.

$$\sum_x \exp\left[N\left(\frac{\beta x^2}{2} - I(x)\right)\right] \asymp \exp\left[N\sup_x\left(\frac{\beta x^2}{2} - I(x)\right)\right].$$

This implies

$$\lim_{N\to\infty} \frac{1}{N} \log Z_{\beta,N} = \sup_{x\in[-1,1]} \left( \frac{\beta x^2}{2} - I(x) \right).$$

The maximum is easy to evaluate: If $\beta \leq 1$ then the maximum is attained at $x = 0$ and the expression is 0. On the other hand, if $\beta > 1$, then there are two maxima inside the open interval $(-1,1)$, namely at $\xi(\beta)$ and $-\xi(\beta)$, and

$$\frac{\beta\xi^2}{2} - I(\xi) = \log\cosh\left(\beta\xi\right) - \frac{\beta\xi^2}{2},$$

by an elementary computation.

There is another nice trick to evaluate the Curie–Weiss free energy: Let $\eta$ be a standard normally distributed random variable, independent of the coin tossings. We denote the probability measure for this variable by $\mathbb{P}$. Then

$$Z_{\beta,N} = E_o \exp \frac{\beta}{2} N S_N^2 = E_o \mathbb{E} \exp \sqrt{\beta N} S_N \eta$$

$$= \mathbb{E} E_o \exp \sqrt{\beta N} S_N \eta = \mathbb{E}\left[ \cosh\left( \frac{\sqrt{\beta}\eta}{\sqrt{N}} \right) \right]^N$$

(6.3)
$$= \frac{1}{\sqrt{2\pi}} \int \exp\left[ -\frac{x^2}{2} \right] \exp\left[ N \log\cosh \frac{\sqrt{\beta}x}{\sqrt{N}} \right] dx$$

$$= \frac{\sqrt{N}}{\sqrt{2\pi}} \int \exp\left[ N\left( \log\cosh\sqrt{\beta}x - \frac{x^2}{2} \right) \right] dx.$$

An elementary analysis of the integral leads to

$$\lim_{N\to\infty} \frac{1}{N} \log Z_{\beta,N} = \sup_{x\in\mathbb{R}} \left[ \log\cosh\sqrt{\beta}x - \frac{x^2}{2} \right].$$

The reader should check that this is the same expression as the one appearing in Theorem 6.1.

To finish this short discussion of the Curie–Weiss model, we consider the properties of the distribution of the mean magnetization $S_N = \frac{1}{N} \sum_{i=1}^{N} \sigma_i$ under the Gibbs measure $P_{\beta,N}$. By symmetry, it is evident that the mean magnetization $E_{\beta,N}(S_N)$ is 0 for all parameters. This, however, is hiding the fact that there is an essential difference between $\beta \leq 1$ and $\beta > 1$. To express this, we better look at the distribution of $S_N$. In fact one has

**Theorem 6.3.** *a) If $\beta \leq 1$ then $S_N$ converges in $P_{\beta,N}-$probability to 0, i.e.*

$$\lim_{N\to\infty} P_{\beta,N}\left( |S_N| \geq \varepsilon \right) = 0$$

*for all $\varepsilon > 0$, or expressed differently, $\lim_{N\to\infty} \mathcal{L}_{P_{\beta,N}}(S_N) = \delta_0$, in the sense of weak convergence of probability measures on $\mathbb{R}$. Here $\mathcal{L}_{P_{\beta,N}}(S_N)$ denotes the law of $S_N$ under the probability distribution $P_{\beta,N}$.*

*b) If $\beta > 1$ then*

$$\lim_{N\to\infty} \mathcal{L}_{P_{\beta,N}}(S_N) = \frac{1}{2}\left(\delta_{-\xi(\beta)} + \delta_{\xi(\beta)}\right),$$

*where $\xi(\beta)$ is the same as in Theorem 6.1.*

**Exercise 6.4.** *Prove this using the analysis in the proof of Theorem 6.1.*

The mean-field equation (6.2) has a very intuitive explanation which is playing a role in the next lecture. Consider any Gibbs measure of the form (6.1) and $i \in \Lambda$. Then, integrating out the $i$-th spin

$$m_i \stackrel{\text{def}}{=} E_{J,\beta,h,\Lambda}(\sigma_i) = E^{(i)}_{J,\beta,h,\Lambda} \tanh\left(h_i + \beta \sum_{k:k\neq i} J_{ik}\sigma_k\right),$$

where $P^{(i)}_{J,\beta,h,\Lambda}$ denotes the distribution on $\Lambda \setminus \{i\}$, dropping all interactions with the $i$-th spin. We introduce a notion which we don't define rigorously, namely that of a "pure state". In the Curie-Weiss model, we have by symmetry $m_i = 0$. However, for $\beta > 1$, the probability space splits essentially into two parts, namely one where $S_N \sim \xi(\beta)$ and one where $S_N \sim -\xi(\beta)$. If we condition on these events, the spins again are close to i.i.d. with mean $\xi(\beta)$ or $-\xi(\beta)$, respectively. We then call these (not really precisely defined) conditional measures the "pure states", and we want now derive $\xi(\beta)$ based on these considerations: We assume that the Gibbs measure can be split (approximately) into such pure states. Inside a "pure state" one would expect that the fluctuations of $\sum_{k:k\neq i} J_{ik}\sigma_k$ are small, so that one can take the expectation on the right-hand side of the above equation under tanh where however $E^{(i)}_{J,\beta,h,\Lambda}$ has to be replaced by one of these pure states. For a large system, there should not be much difference between the system on $\Lambda$ and that on $\Lambda \setminus \{i\}$, i.e. the expectation should be $\approx m_k$ again. Therefore, we get

$$m_i \approx \tanh\left(h_i + \beta \sum_{k:k\neq i} J_{ik}m_k\right)$$

in a large system. If the $h_i = 0$ and $J_{ik} = \frac{1}{N}$, one expects that the $m_i$ are all the same, so we get (6.2) for $\xi = m_i$.

## 6.3. The Sherrington–Kirkpatrick Model: High Temperature and No Magnetic Field

The Sherrington–Kirkpatrick model has again $\Lambda_N = \{1, \ldots, N\}$ and is given by

$$(6.4) \qquad P_{J,\beta,h,N}(\sigma) = \frac{1}{Z_{J,\beta,h,N}} 2^{-N} \exp\left[\frac{1}{\sqrt{N}} \sum_{1\leq i<j\leq N} J_{i,j}\sigma_i\sigma_j + h\sum_{i=1}^{N}\sigma_i\right].$$

Here $h$ is just a real parameter. However, the $J_{i,j}$ are i.i.d. random variables, with mean 0 and variance 1. We denote the probability space on which they are defined by $(\Omega, \mathcal{F}, \mathbb{P})$. For most of the theory, the exact distribution of the $J_{i,j}$ is not of much importance. The simplest choice however is to take the $J_{i,j}$ standard normally distributed, which we do through the rest of these lectures. There are several variations, for instance by taking the parameter $h$ also random. This does not lead to any substantially new effects, so we keep with the above situation. One might wonder about the normalizing factor $1/\sqrt{N}$ in the interaction term, which is different from the one in the Curie–Weiss case which is $1/N$. However, a moment's reflection reveals that this is the proper way to do it. $\sigma_i$ interacts with the other spin variables through $\frac{1}{\sqrt{N}}(\sum_{j:j>i} J_{ij}\sigma_j + \sum_{j:j<i} J_{ji}\sigma_j)$. This is a random variable with mean 0 and variance $(N-1)/N \sim 1$. Therefore, the total interaction of $\sigma_i$ with the other spin variables is typically of order 1, as in the Curie–Weiss or the Ising case. In this section, we consider the case $h = 0$, and in the next lecture the more delicate case where $h \neq 0$.

Most of the statements one would like to make have to be formulated either as statements which hold almost surely with respect to $\mathbb{P}$, i.e. the realization of the $J$ variables, or by taking expectations with respect to these variables.

Unfortunately, it is not even known that the free energy $f(\beta, h) = \lim_{N\to\infty} \frac{1}{N} \log Z_{J,\beta,h,N}$ exists almost surely and for all parameters $\beta$. It is however known that if it exists, it is nonrandom and given by

$$(6.5) \qquad f(\beta, h) = \lim_{N\to\infty} \frac{1}{N} \mathbb{E} \log Z_{J,\beta,h,N}.$$

This property is called "self-averaging" of the free energy. We shall prove the following result ([1], [13], [23], [69]):

**Theorem 6.5.** *If $h = 0$ and $\beta \leq 1$, then the free energy exists almost surely and equals*

$$f(\beta) = \frac{\beta^2}{4}.$$

**Remark 6.6.** *By Jensen's inequality and (6.5) we see that*

$$(6.6) \qquad f(\beta) \leq \lim_{N\to\infty} \frac{1}{N} \log \mathbb{E} Z_{J,\beta,\Lambda}$$

$$= \lim_{N\to\infty} \frac{1}{N} \log E_o \exp\left[ \frac{\beta^2}{2N} \sum_{1 \leq i < j \leq N} \sigma_i^2 \sigma_j^2 \right] = \beta^2/4.$$

*It is known that this is a strict inequality for $\beta > 1$. This has been proved in [11].*

**Proof of Theorem 6.5** We assume $\beta < 1$. The case $\beta = 1$ can then be proved by taking the limit. The main task is to prove that $\log \mathbb{E}(Z_{J,\beta,N}) \approx \mathbb{E}(\log Z_{J,\beta,N}) \approx \log Z_{J,\beta,N}$. The last approximation is actually easy (and holds true for all $\beta$).

We use the so called second moment method to prove the first approximation. For this, we have to calculate the second moment of $Z$ :

$$
\mathbb{E}((Z_{J,\beta,N})^2) = \mathbb{E}\Big(\Big(E_o \exp\Big[\frac{\beta}{\sqrt{N}} \sum_{1\leq i<j\leq N} J_{i,j}\sigma_i\sigma_j\Big]\Big)^2\Big)
$$

$$
= \mathbb{E}\Big(E_o^{(2)} \exp\Big[\frac{\beta}{\sqrt{N}}\Big(\sum_{1\leq i<j\leq N} J_{i,j}\big(\sigma_i\sigma_j + \sigma_i'\sigma_j'\big)\Big)\Big]\Big)
$$

(6.7)
$$
= E_o^{(2)} \exp\Big[\frac{\beta^2}{2N} \sum_{1\leq i<j\leq N} (\sigma_i\sigma_j + \sigma_i'\sigma_j')^2\Big]
$$

$$
= \exp\Big[\frac{\beta^2(N-1)}{2}\Big] \times E_o^{(2)} \exp\Big[\frac{\beta^2}{N} \sum_{1\leq i<j\leq N} \sigma_i\sigma_j\sigma_i'\sigma_j'\Big].
$$

In the second equation, we use a trick, that will recur again and again: We write the square of an expectation in terms of an expectation of the product of two independent copies of the set of random variables, in our case denoted by $\sigma, \sigma'$. We then also write $P^{(2)}$ for the probability measure governing these two independent copies, and more generally, with $k$ "replicas", we write $P^{(k)}$. Remark that $\exp[\frac{\beta^2(N-1)}{2}] = (\mathbb{E}Z_{J,\beta,N})^2$. We want to prove that the second factor on the right hand side of (6.7) is bounded by a constant as soon as $\beta < 1$. Now, if $(\sigma_i)$ and $(\sigma_i')$ are two independent coin tossing sequences, then also the sequence $(\sigma_i\sigma_i')_{1\leq i\leq N}$ is a (simple) coin tossing sequence, and therefore

$$
E_o^{(2)} \exp\Big[\frac{\beta^2}{N} \sum_{1\leq i<j\leq N} \sigma_i\sigma_j\sigma_i'\sigma_j'\Big] = E_o\Big(\exp\Big[\frac{\beta^2}{N} \sum_{1\leq i<j\leq N} \sigma_i\sigma_j\Big]\Big)
$$

$$
= \exp\Big[-\frac{\beta^2}{2}\Big] E_o\Big(\exp\Big[\frac{\beta^2 N}{2} S_N^2\Big]\Big).
$$

The second factor is exactly the one we had encountered in the Curie–Weiss model, but we want to estimate that slightly more carefully than in the last section (when $\beta < 1$). According to the second proof in the last section, the expectation on the right hand side can be written as

$$
\frac{\sqrt{N}}{\sqrt{2\pi}} \int \exp\Big[N\Big(\log\cosh\beta x - \frac{x^2}{2}\Big)\Big] dx.
$$

We now use the elementary inequality $\log\cosh(x) \leq x^2/2$. Therefore

$$
E_o\Big(\exp\Big[\frac{\beta^2 N}{2} S_N^2\Big]\Big) \leq \frac{1}{\sqrt{2\pi}} \int \exp\Big[-\frac{x^2}{2}(1-\beta^2)\Big] dx = \frac{1}{\sqrt{1-\beta^2}},
$$

for $\beta < 1$. Therefore, we have proved that

(6.8)
$$
\mathbb{E}((Z_{J,\beta,N})^2) \leq c(\beta)(\mathbb{E}(Z_{J,\beta,N}))^2,
$$

in this case.

To finish the proof of the theorem, we have to use the so called Gaussian isoperimetric inequality. I will not prove it here, as it is outside the scope of these lectures, referring for a proof e.g. to the St. Flour Lecture Notes of Ledoux [41].

**Theorem 6.7.** *Consider the standard normal distribution* $\mu$ *on* $\mathbb{R}^K$. *If* $f : \mathbb{R}^K \to \mathbb{R}$ *is Lipshitz continuous, with Lipshitz constant* $L$ *(i.e.* $|f(x) - f(y)| \leq L\|x - y\|$, *where* $\|\cdot\|$ *is the Euclidean norm on* $\mathbb{R}^K$), *then*

$$\mu(\{\, x : |f(x) - \int f \, d\mu| \geq t \,\}) \leq 2\exp\left[-\frac{t^2}{2L^2}\right].$$

We apply this to $f_N : \mathbb{R}^{N(N-1)/2} \to \mathbb{R}$ given by

$$f_N((x_{i,j})_{1\leq i<j\leq N}) = \frac{1}{N}\log 2^{-N}\sum_\sigma \exp\left[\frac{\beta}{\sqrt{N}}\sum_{1\leq i<j\leq N}x_{ij}\sigma_i\sigma_j\right].$$

Then

$$|f_N(x) - f_N(y)| \leq \frac{\beta}{N^{3/2}}\sum_{1\leq i<j\leq N}|x_{ij} - y_{ij}| \leq \frac{\beta}{\sqrt{N}}\|x - y\|,$$

by Cauchy-Schwartz. Therefore,

$$(6.9) \qquad \mathbb{P}\left(\left|\frac{1}{N}\log Z_N - \frac{1}{N}\mathbb{E}\log Z_N\right| \geq \frac{t}{\sqrt{N}}\right) \leq \exp\left[-\frac{t^2}{2\beta^2}\right].$$

With this estimate, we can now easily finish the proof of Theorem 6.5.

Define $A_N \overset{\text{def}}{=} \{Z_N \geq \mathbb{E}Z_N/2\} = \{Z_N \geq \frac{1}{2}\exp[\frac{\beta^2(N-1)}{4}]\}$. Then

$$\mathbb{E}Z_N = \mathbb{E}(Z_N; A_N^c) + \mathbb{E}(Z_N; A_N) \leq \mathbb{E}Z_N/2 + \sqrt{\mathbb{E}(Z_N^2)\,\mathbb{P}(A_N)},$$

and therefore

$$\mathbb{P}(A_N) \geq \frac{\mathbb{E}(Z_N)^2}{4\mathbb{E}(Z_N^2)} \geq c(\beta) > 0$$

if $\beta < 1$ by (6.8). Using this together with the above concentration inequality (6.9), one sees that

$$\frac{1}{N}\mathbb{E}\log Z_N \geq \frac{1}{N}\log\frac{\mathbb{E}Z_N}{2} - O\left(\frac{1}{\sqrt{N}}\right)$$

$$= \frac{1}{N}\log\mathbb{E}Z_N - O\left(\frac{1}{\sqrt{N}}\right) = \frac{\beta^2}{4} - O\left(\frac{1}{\sqrt{N}}\right).$$

Therefore, it follows that

$$\liminf_{N\to\infty}\frac{1}{N}\mathbb{E}\log Z_N \geq \frac{\beta^2}{4}.$$

On the other hand one has by Jensens inequality $\mathbb{E}\log Z_N \leq \log\mathbb{E}Z_N$, and therefore, one gets

$$(6.10) \qquad \lim_{N\to\infty}\frac{1}{N}\mathbb{E}\log Z_N = \frac{\beta^2}{4}.$$

Using the concentration inequality (6.6) again, and Borel-Cantelli, one gets

$$\lim_{N \to \infty} \frac{1}{N} \log Z_N = \frac{\beta^2}{4}, \quad \text{almost surely},$$

finishing the proof of Theorem 6.5.

It is quite a miracle that the second moment method works for the SK-model up to the correct critical temperature parameter $\beta_c = 1$. This is not the case even in the most simple caricature of a spin glass model, the random energy model, which will be discussed in Lecture 8.

**Remark 6.8.** *One actually knows that for $\beta < 1$*

$$|\mathbb{E} \log Z_N - \log \mathbb{E} Z_N| = O(1),$$

*and furthermore* $\log Z_N - \frac{\beta^2}{4} N$ *is asymptotically normally distributed (see* [1], [13]*).*

# Lecture 7:

# The Sherrington-Kirkpatrick Model:
# High Temperature and Nonzero Magnetic Field

We consider the SK-model (6.4) for $h \neq 0$ and small $\beta$. There are good reasons to be interested in this case. The high temperature $h = 0$ case is in fact deceptively simple, mainly due to the inherent symmetry properties.

As soon as $h \neq 0$, the free energy is for no $\beta > 0$ equal to the annealed free energy $\lim_{N \to \infty} \frac{1}{N} \log \mathbb{E} Z_{J,\beta,h,N}$ which is of course easy to evaluate. In order to state the result, we consider the following equation for $q = q(\beta, h) > 0$:

$$(7.1) \qquad q = \frac{1}{\sqrt{2\pi}} \int \tanh^2 \left( h + \beta \sqrt{q} x \right) \exp \left[ -\frac{x^2}{2} \right] dx.$$

In the physics literature, it is taken for granted that for $h > 0$, this fixed-point equation has a unique positive solution, but there seems to exist no (published) proof. Plotting the graph of the function on the right-hand side on a computer, the claim looks being correct.

For $h = 0$, there is of course always the solution $q = 0$, but it becomes unstable for $\beta > 1$ where two other (symmetric) solutions pop up.

**Exercise 7.1.** *There exists $\beta_0 > 0$ such that for $\beta \leq \beta_0$ and all $h > 0$, there exists a unique positive solution $q = q(\beta, h)$ of (7.1).*

**Theorem 7.2.** ([69]) *There exists $\beta_0 > 0$ such that for $\beta \leq \beta_0$ and all $h > 0$, the free energy exists and is given by*

$$f(\beta, h) = \frac{\beta^2}{4} (1 - q)^2 + \frac{1}{\sqrt{2\pi}} \int \log \cosh \left( h + \beta \sqrt{q} x \right) \exp \left[ -\frac{x^2}{2} \right] dx,$$

*where $q = q(\beta, h)$.*

In physics literature, it is claimed that the above formula is correct as long as the following condition due to de Almeida and Thouless [14] is satisfied:

$$\beta^2 \frac{1}{\sqrt{2\pi}} \int \frac{1}{\cosh^4 (h + \beta \sqrt{q} x)} \exp \left[ -\frac{x^2}{2} \right] dx < 1.$$

The proof of Theorem 7.2 I am giving below is essentially the one by Talagrand [69], with some modifications.

The high temperature region is characterized by the property that different spins are under the Gibbs measure essentially uncorrelated in a sense to be made precise below. One feature which makes things for $h \neq 0$ quite more complicated is the fact that whereas the spins in the $h = 0$ case are evidently symmetric, and therefore have expectation 0, this is no longer the case here. We usually suppress $h$ and $N$ in the notation if there is no danger of confusion, but they should always be remembered to be present. We write $P = P_{J,\beta,h,N}$ for the Gibbs measure, and define

$$m_i = E\sigma_i.$$

The reader should keep in mind that this is still a random variable, depending on the $J$ variables, and of course, these quantities depend also on $\beta, h$ and $N$. A quantity which naturally measures correlations between different spins is given by

$$\gamma_N(\beta) \overset{\text{def}}{=} \mathbb{E}\left(\text{cov}_P(\sigma_1, \sigma_2)^2\right)$$

**Proposition 7.3.**

    a) *There exists $\beta_0 > 0$ and $\rho < 1$ such that for $\beta \leq \beta_o$, we have*

$$\gamma_{N+2}(\beta\sqrt{1 + 2/N}) \leq \rho\gamma_N(\beta) + \frac{C}{N}.$$

    b) $\hspace{5cm} \gamma_N(\beta) \leq \frac{C}{N}.$

*Proof.* b) follows easily from a). We leave this as an exercise to the reader. So we prove a).

The idea of the proof is to consider the system of size $N$, with its Gibbs measure $P_{\beta,N}$, and then write out expectations with respect to the larger system of size $N + 2$ by splitting the interaction between the "old" spins $\sigma = (\sigma_1, \ldots, \sigma_N)$ and the two "new" ones $\tau = (\tau_1, \tau_2) \overset{\text{def}}{=} (\sigma_{N+1}, \sigma_{N+2})$. We want to keep the prefactor $\beta/\sqrt{N}$ also for the $(N + 2)$-system, which means that we have to change the temperature parameter to $\beta\sqrt{1 + 2/N}$ for the $(N+2)$-system. Then the interaction between $\sigma$ and $\tau$ is given by $\zeta(\sigma, \tau) \overset{\text{def}}{=} \frac{\beta}{\sqrt{N}}\sum_{i=1}^{N}\sigma_i(J_{i,N+1}\tau_1 + J_{i,N+2}\tau_2)$, We will also write $\mathbb{E}_N$ for taking expectation with respect to the $J_{ij}$-variables $1 \leq i < j \leq N$ only, and $\mathbb{E}_{\text{cav}}$ for the expectation with respect to the variables $J_{i,N+1}$ and $J_{i,N+2}$, $1 \leq i \leq N$. (The index "cav" stands for "cavity". The method used by Talagrand is based on ideas which in the physics literature are called "cavity method"). We then get

$$\text{cov}_{P_{\beta\sqrt{1+2/N},N+2}}(\sigma_{N+1}, \sigma_{N+2}) = \text{cov}_{P_{\beta\sqrt{1+2/N},N+2}}(\tau_1, \tau_2)$$

$$= \frac{\sum_\tau E_{\beta,N}(\overline{\tau}_1\overline{\tau}_2\exp[\zeta(\sigma,\tau) + \frac{\beta}{\sqrt{N}}J_{N+1,N+2}\tau_1\tau_2 + h(\tau_1 + \tau_2)])}{\sum_\tau E_{\beta,N}(\exp[\zeta(\sigma,\tau) + \frac{\beta}{\sqrt{N}}J_{N+1,N+2}\tau_1\tau_2 + h(\tau_1 + \tau_2)])},$$

where $\overline{\tau}_i = \tau_i - m_i$. Writing

$$(7.2) \qquad A(\tau) = E_{\beta,N}\left(\exp\left[\zeta(\sigma,\tau) + \frac{\beta}{\sqrt{N}}J_{N+1,N+2}\tau_1\tau_2 + h(\tau_1 + \tau_2)\right]\right),$$

and a simple computation yields

$$\text{cov}_{P_{\beta\sqrt{1+2/N},N+2}}(\tau_1, \tau_2) = 4\frac{A(1,1)A(-1,-1) - A(1,-1)A(-1,1)}{(\sum_\tau A(\tau))^2}.$$

By Jensen's inequality, we get $\sum_\tau A(\tau) \geq 4$, and therefore

$$(7.3) \quad \gamma_{N+2}(\beta\sqrt{1+2/N}) \leq C\mathbb{E}([A(1,1)A(-1,-1) - A(1,-1)A(-1,1)]^2).$$

We will of course always assume that $\beta$ is bounded by some fixed constant, say $\beta \leq 1$. Remark also that the $\mathbb{E}$-expectations of arbitrary exponents of $A$-variables stay bounded. Using this, it is easily checked that integrating out $J_{N+1,N+2}$ in the above expression leads to factors of the form $(1 + O(1/N))$. This is a general feature of mean-field models: The direct interaction between two spins is never of any importance for their correlations. We therefore can as well leave the $J_{N+1,N+2}$-terms out, catching an error of order $1/N$. We therefore define

$$A_o(\tau) = E_{\beta,N}(\exp[\zeta(\sigma,\tau)]),$$

and then get the same estimate as (7.3), but with $A$ replaced by $A_o$ and an additional summand $C/N$. Furthermore,

$$A_o(1,1)A_o(-1,-1) - A_o(1,-1)A_o(-1,1)$$
$$= E_{\beta,N}^{(2)}\left(\exp\left[\frac{\beta}{\sqrt{N}}\sum_{i=1}^{N}(\sigma_i - \sigma_i')(J_{i,N+1} + J_{i,N+2})\right]\right)$$
$$- E_{\beta,N}^{(2)}\left(\exp\left[\frac{\beta}{\sqrt{N}}\sum_{i=1}^{N}(\sigma_i - \sigma_i')(J_{i,N+1} - J_{i,N+2})\right]\right),$$

where $E_{\beta,N}^{(2)}$ is an expectation with respect to two independent copies $\sigma, \sigma'$. Remark that $J_{i,N+1} + J_{i,N+2}$ and $J_{i,N+1} - J_{i,N+2}$ are independent. Therefore

$$\mathbb{E}_{cav}([A_o(1,1)A_o(-1,-1) - A_o(1,-1)A_o(-1,1)]^2)$$
$$(7.4) \quad = 2\text{var}_{cav}\left(E_{\beta,N}^{(2)}\left(\exp\left[\frac{\beta}{\sqrt{N}}\sum_{i=1}^{N}(\sigma_i - \sigma_i')(J_{i,N+1} + J_{i,N+2})\right]\right)\right)$$
$$= 2E_{\beta,N}^{(4)}\left(e^{\frac{\beta^2}{N}\|\sigma^{(1)}-\sigma^{(2)}+\sigma^{(3)}-\sigma^{(4)}\|^2} - e^{\frac{\beta^2}{N}(\|\sigma^{(1)}-\sigma^{(2)}\|^2+\|\sigma^{(3)}-\sigma^{(4)}\|^2)}\right)$$
$$= 2E_{\beta,N}^{(4)}\left(e^{\frac{\beta^2}{N}(\|\sigma^{(1)}-\sigma^{(2)}\|^2+\|\sigma^{(3)}-\sigma^{(4)}\|^2)}\right)\left(e^{\frac{2\beta^2}{N}\langle\sigma^{(1)}-\sigma^{(2)},\sigma^{(3)}-\sigma^{(4)}\rangle} - 1\right)$$

where $\|\sigma\|^2 \stackrel{\text{def}}{=} \sum_i \sigma_i^2$ and $\langle\sigma,\sigma'\rangle = \sum_i \sigma_i\sigma_i'$. Interchanging the role of $\sigma^{(1)}$ and $\sigma^{(2)}$, we see that the above expression is

$$2E_{\beta,N}^{(4)}\left(e^{\frac{\beta^2}{N}(\|\sigma^{(1)}-\sigma^{(2)}\|^2+\|\sigma^{(3)}-\sigma^{(4)}\|^2)}\right)\left(\cosh\left(\frac{2\beta^2}{N}\langle\sigma^{(1)}-\sigma^{(2)},\sigma^{(3)}-\sigma^{(4)}\rangle\right) - 1\right)$$
$$\leq \frac{C\beta^4}{N^2}E_{\beta,N}^{(4)}(\langle\sigma^{(1)}-\sigma^{(2)},\sigma^{(3)}-\sigma^{(4)}\rangle^2).$$

We now integrate out also the $J_{ij}$, $1 \le i < j \le N$:

$$
\mathbb{E}_N E_{\beta,N}^{(4)} \left[ \left( \frac{1}{N^2} \langle \sigma^{(1)} - \sigma^{(2)}, \sigma^{(3)} - \sigma^{(4)} \rangle^2 \right) \right]
$$

(7.5)
$$
= \frac{N(N-1)}{N^2} \mathbb{E}_N E_{\beta,N}^{(4)} \left[ \left( \sigma_1^{(1)} - \sigma_1^{(2)} \right) \left( \sigma_1^{(3)} - \sigma_1^{(4)} \right) \right.
$$
$$
\left. \times \left( \sigma_2^{(1)} - \sigma_2^{(2)} \right) \left( \sigma_2^{(3)} - \sigma_2^{(4)} \right) \right] + O\left( \frac{1}{N} \right)
$$
$$
= \mathbb{E}_N \left[ E_{\beta,N}^{(2)} \left( \left( \sigma_1^{(1)} - \sigma_1^{(2)} \right) \left( \sigma_2^{(1)} - \sigma_2^{(2)} \right) \right) \right]^2 + O\left( \frac{1}{N} \right)
$$
$$
= \gamma_N(\beta) + O\left( \frac{1}{N} \right).
$$

Combining (7.3)-(7.5), this finishes the proof of Proposition 7.3.          □

The above proposition tells us that for small enough $\beta > 0$, the spin variables $\sigma_i$ are essentially uncorrelated (and therefore independent, because they take only two values), with large $\mathbb{P}$-probability. We now want to derive from this that also the $m_i$ are essentially uncorrelated. A crucial property is that the $m_i$ satisfy (approximately) the so-called TAP-equation (see [71]):

$$
m_i \approx \tanh\left( h + \frac{\beta}{\sqrt{N}} \sum_{j: j \ne i} J_{ij} m_j^{(i)} \right),
$$

where $m_j^{(i)}$ is the mean of $\sigma_j$ where we suppress the interaction of the $i$-the spin with the rest. We make this precise below. To start with, we prove an approximation in a setup which is slightly more general than what is actually needed.

Let $P$ be a probability measure on $\Sigma_N \overset{\text{def}}{=} \{-1, 1\}^N$ with $m_i \overset{\text{def}}{=} E\sigma_i$, and $\gamma_{ij} \overset{\text{def}}{=} \text{cov}_P(\sigma_i, \sigma_j)$. Let further $\xi_1, \xi_2, \ldots, \xi_N$ be i.i.d. standard normally distributed random variables defined on $(\Omega, \mathcal{F}, \mathbb{P})$. We will use the abbreviation

$$
Y_k^{(\beta,N)} \overset{\text{def}}{=} \frac{\beta}{\sqrt{N}} \sum_{i=1}^k \sigma_i \xi_i,
$$

but we will usually drop the $\beta$- and $N$-dependence in the notation, and just write $Y_k$.

In the discussion below, it is convenient if the $\xi_i$-variables are all not too large. We can achieve this by truncating the $\xi$-variables slightly, putting $\overline{\xi_i} \overset{\text{def}}{=} \xi_i I$ ($|\xi_i| \le N^{1/4}$). Of course, the probability that *any* of the $\xi$-variables differs from the $\overline{\xi}$-variables is decaying faster than any polynomial, and this event will therefore be negligible in the end (in Proposition 7.5). We can therefore assume $|\xi_i| \le N^{1/4}$ for all $i \le N$.

We denote by $\mathcal{F}_k$, $k \le N$ the $\sigma$-field generated by $\xi_1, \ldots, \xi_k$. We write $\Delta_k$ for a generic $\mathcal{F}_k$-measurable random variable, satisfying $\mathbb{E}\Delta_k^2 \le C$, not necessarily the same at different occurrences, where $C$ may depend on $\beta$ and $h$, but stays bounded for $\beta, h$ in bounded regions.

Let

$$\Gamma \stackrel{\text{def}}{=} \sqrt{\frac{1}{N^2} \sum_{i,j=1}^{N} \gamma_{ij}^2}, \quad \Gamma_k = \Gamma + \frac{1}{N} \sum_{j=1}^{k-1} \gamma_{jk}^2.$$

**Lemma 7.4.** $|E((\sigma_k - m_k)\cosh(h + Y_{k-1}))|$ and $|E(\sigma_k - m_k)\sinh(h + Y_{k-1}))|$ are bounded by $\sqrt{\Gamma_k}\Delta_{k-1}$.

*Proof.* It suffices to consider

$$\eta_k \stackrel{\text{def}}{=} E((\sigma_k - m_k)e^{h+Y_{k-1}}),$$

and then we let drop $h$ out, as it enters only into the constant. By the usual representation, we have (for $h = 0$)

$$\mathbb{E}\eta_k^2 = E^{(4)}\left( (\sigma_k^{(1)} - \sigma_k^{(2)})(\sigma_k^{(3)} - \sigma_k^{(4)}) \mathbb{E}\left( \exp\left[ \frac{\beta}{\sqrt{N}} \sum_{j=1}^{k-1} \xi_j (\sigma_j^{(1)} + \sigma_j^{(3)}) \right] \right) \right)$$

$$= C_{k,N} E^{(4)}\left( (\sigma_k^{(1)} - \sigma_k^{(2)})(\sigma_k^{(3)} - \sigma_k^{(4)}) \exp\left[ \frac{\beta^2}{N} \sum_{j=1}^{k-1} (\sigma_j^{(1)}\sigma_j^{(3)} - m_j^2) \right] \right),$$

where the $C_{k,N}$ stay uniformly bounded. We now write $e^x = 1 + x + (e^x - 1 - x)$, where the third summand is bounded by $Cx^2$ for $x$-values in a bounded region. As $E^{(4)}(\sigma_k^{(1)} - \sigma_k^{(2)})(\sigma_k^{(3)} - \sigma_k^{(4)}) = 0$, we get

$$\mathbb{E}\eta_k^2 \leq CE^{(4)}\left\{ (\sigma_k^{(1)} - \sigma_k^{(2)})(\sigma_k^{(3)} - \sigma_k^{(4)}) \frac{1}{N} \sum_{j=1}^{k-1} (\sigma_j^{(1)}\sigma_j^{(3)} - m_j^2) \right\}$$

$$+ CE^{(4)}\left( \frac{1}{N} \sum_{j=1}^{k-1} (\sigma_j^{(1)}\sigma_j^{(3)} - m_j^2) \right)^2$$

$$\leq C\Gamma_k$$

As $\eta_k$ is evidently $\mathcal{F}_{k-1}$-measurable, the lemma is proved.    □

**Proposition 7.5.**

$$\frac{E\sinh(h + Y_N)}{E\cosh(h + Y_N)} = \tanh\left( h + \frac{\beta}{\sqrt{N}} \sum_{j=1}^{N} \xi_j m_j \right) + R,$$

with

(7.6)                    $$\mathbb{E}R^2 \leq C\left( \Gamma + \frac{1}{\sqrt{N}} \right).$$

*Proof.* Let

$$\phi_k \stackrel{\text{def}}{=} \frac{E\sinh(h + Y_k)}{E\cosh(h + Y_k)}.$$

We also use the abbreviations $S_k \stackrel{\text{def}}{=} \sinh(h+Y_k)$, $C_k \stackrel{\text{def}}{=} \cosh(h+Y_k)$. Expanding $S_k$ and $C_k$ in $\xi_k \sigma_k / \sqrt{N}$ we obtain.

$$\phi_k = \frac{ES_{k-1} + \frac{\xi_k}{\sqrt{N}} E(\sigma_k C_{k-1}) + \frac{\xi_k^2}{2N} ES_{k-1} + N^{-3/2}\Delta_k}{EC_{k-1} + \frac{\xi_k}{\sqrt{N}} E(\sigma_k S_{k-1}) + \frac{\xi_k^2}{2N} EC_{k-1} + N^{-3/2}\Delta'_k}$$

$$= \frac{ES_{k-1} + \frac{\xi_k}{\sqrt{N}} E(\sigma_k C_{k-1})}{EC_{k-1} + \frac{\xi_k}{\sqrt{N}} E(\sigma_k S_{k-1})} + N^{-3/2}\Delta_k,$$

the latter because we assumed $|\xi_k| \leq N^{1/4}$, and $|E(\sigma_k S_{k-1})| \leq EC_{k-1}$, so that the denominator stays away from 0. With the abbreviation $\varepsilon_k \stackrel{\text{def}}{=} \xi_k m_k / \sqrt{N}$ we get

$$\phi_k = \frac{\phi_{k-1} + \varepsilon_k + \frac{\xi_k}{\sqrt{N}} \psi_{k-1}}{1 + \varepsilon_k \phi_{k-1} + \frac{\xi_k}{\sqrt{N}} \eta_{k-1}} + N^{-3/2}\Delta_k,$$

where

$$\psi_{k-1} \stackrel{\text{def}}{=} \frac{E((\sigma_k - m_k)C_{k-1})}{EC_{k-1}}, \; \eta_{k-1} \stackrel{\text{def}}{=} \frac{E((\sigma_k - m_k)S_{k-1})}{EC_{k-1}}.$$

Remark that these variables are $\mathcal{F}_{k-1}$-measurable. Expanding the quotient, we get
(7.7)
$$\phi_k = \frac{\phi_{k-1} + \varepsilon_k}{1 + \varepsilon_k \phi_{k-1}} + \frac{\xi_k}{\sqrt{N}} \left( \psi_{k-1} - \eta_{k-1}\phi_{k-1} \right) + \frac{\xi_k^2}{N} \Delta_{k-1} \left( \eta_{k-1} + \psi_{k-1} \right) + N^{-3/2}\Delta_k.$$

Put now $\phi_k^0 \stackrel{\text{def}}{=} \tanh(h + \frac{1}{\sqrt{N}} \sum_{j=1}^k \xi_j m_j)$, and set $\phi_k = \phi_k^0 + \delta_k$. The crucial point is that if we can replace $\phi_{k-1}$ by $\phi_{k-1}^0$, then the first summand on the right hand side of (7.7) is just $\phi_k^0$, up to an error of order $\varepsilon_k^3 = N^{-3/2}\Delta_k$. After some regrouping, we get

$$\frac{\phi_{k-1}^0 + \delta_{k-1} + \varepsilon_k}{1 + \varepsilon_k(\phi_{k-1}^0 + \delta_{k-1})} = \frac{\phi_{k-1}^0 + \varepsilon_k}{1 + \varepsilon_k \phi_{k-1}^0} + \delta_{k-1}[1 + \varepsilon_k \Delta_{k-1} + \varepsilon_k^2 \Delta'_{k-1}] + N^{-3/2}\Delta''_k$$

$$= \phi_k^0 + \delta_{k-1}[1 + \varepsilon_k \Delta_{k-1} + \varepsilon_k^2 \Delta'_{k-1}] + N^{-3/2}\Delta''_k,$$

and implementing into (7.7) we have

$$\delta_k = \delta_{k-1} + \frac{\xi_k}{\sqrt{N}} \left[ \psi_{k-1} - \eta_{k-1}\phi_{k-1} + \delta_{k-1}\Delta_{k-1} \right]$$

$$+ \frac{\xi_k^2}{N} \left[ \Delta'_{k-1} \left( \eta_{k-1} + \psi_{k-1} \right) + \delta_{k-1}\Delta''_{k-1} \right] + N^{-3/2}\Delta'''_k,$$

leading with the help of Lemma 7.4 to

$$\mathbb{E}\delta_k^2 \le \mathbb{E}\delta_{k-1}^2\left(1 + \frac{C}{N}\right) + \frac{C}{N}\mathbb{E}\left(|\delta_{k-1}|\,|\eta_{k-1} + \psi_{k-1}|\right) + \frac{C}{N}\Gamma_k + CN^{-3/2}$$
$$\le \mathbb{E}\delta_{k-1}^2\left(1 + \frac{C}{N}\right) + \frac{C}{N}\sqrt{\mathbb{E}\delta_{k-1}^2\,\mathbb{E}\left(\eta_{k-1} + \psi_{k-1}\right)^2} + \frac{C}{N}\Gamma_k + CN^{-3/2}$$
$$\le \mathbb{E}\delta_{k-1}^2\left(1 + \frac{C}{N}\right) + \frac{C}{N}\Gamma_k + CN^{-3/2}.$$

From this inequality, the estimate (7.6) follows by iteration.    □

**Remark 7.6.** *We will need a slight extension of the above result. Consider the same* $P$, *but now two independent sets of normally distributed random variables* $\xi_i^{(1)}$ *and* $\xi_i^{(2)}$. *Let furthermore*

$$Y_k^{(i)} \stackrel{\text{def}}{=} \frac{\beta}{\sqrt{N}}\sum_{j=1}^k \xi_j^{(i)}\sigma_i.$$

*Then*

$$\frac{E(\sinh(h + Y_N^{(1)})\cosh(h + Y_N^{(2)}))}{E(\cosh(h + Y_N^{(1)})\cosh(h + Y_N^{(2)}))} = \tanh\left(h + \frac{\beta}{\sqrt{N}}\sum_{j=1}^N \xi_j^{(1)}m_j\right) + R,$$

*with* $R$ *satisfying the same estimate (7.6).*

*Proof.* This is the same argument with some slight modifications as the one given above. We put here

$$\phi_k \stackrel{\text{def}}{=} \frac{E(\sinh(h + Y_k^{(1)})\cosh(h + Y_N^{(2)}))}{E(\cosh(h + Y_k^{(1)})\cosh(h + Y_N^{(2)}))},$$

and remark that $\phi_0 = \tanh(h)$. Expanding as above, we get

$$\frac{E(\sinh(h + Y_k^{(1)})\cosh(h + Y_N^{(2)}))}{E(\cosh(h + Y_k^{(1)})\cosh(h + Y_N^{(2)}))} = \frac{E(S_{k-1}^{(1)}C_N^{(2)}) + \frac{\xi_k}{\sqrt{N}}E(\sigma_k C_{k-1}^{(1)}C_N^{(2)})}{E(C_{k-1}^{(1)}C_N^{(2)}) + \frac{\xi_k}{\sqrt{N}}E(\sigma_k S_{k-1}^{(1)}C_N^{(2)})} + N^{-3/2}\Delta_N,$$

and then the rest runs as above, we only have to generalize the Lemma 7.4 to the estimates

$$\left|E\left((\sigma_k - m_k)\cosh(h + Y_{k-1}^{(1)})\right)\cosh(h + Y_N^{(2)})\right| \le \sqrt{\Gamma_k}\Delta_{k-1}^{(1)}\Delta_N^{(2)},$$

and the same for $|E((\sigma_k - m_k)\sinh(h + Y_{k-1}^{(1)}))\cosh(h + Y_N^{(2)})|$.    □

The idea of the proof of Theorem 7.2 consists in differentiating the (finite $N$) free energy with respect to $\beta$ and control what happens in the $N \to \infty$ limit.

Let

$$F_N(\beta, h) \stackrel{\text{def}}{=} \log Z_{\beta,h,N}.$$

Then

$$\frac{\partial F_N}{\partial \beta}(\beta, h) = \frac{1}{\sqrt{N}} \sum_{1 \leq i < j \leq N} J_{i,j} E_{\beta,h,N}(\sigma_i \sigma_j),$$

and therefore for $\beta \leq \beta_o$, $\beta_o$ small enough:

$$\mathbb{E}\frac{1}{N}\frac{\partial F_N}{\partial \beta}(\beta, h) = \mathbb{E}\frac{1}{N^{3/2}} \sum_{1 \leq i < j \leq N} J_{i,j} E_{\beta,h,N}(\sigma_i \sigma_j)$$

$$= \frac{\beta}{N^2} \mathbb{E} \sum_{1 \leq i < j \leq N} \left(1 - [E_{\beta,h,N}(\sigma_i \sigma_j)]^2\right)$$

$$= \frac{\beta}{2}\left(1 - \mathbb{E}[E_{\beta,h,N}(\sigma_1 \sigma_2)]^2\right) + O\left(\frac{1}{N}\right)$$

$$= \frac{\beta}{2}\left(1 - \mathbb{E}[\mathrm{cov}\, P_{\beta,h,N}(\sigma_1 \sigma_2) + m_1 m_2]^2\right) + O\left(\frac{1}{N}\right)$$

$$= \frac{\beta}{2}\left(1 - \mathbb{E}[m_1^2 m_2^2] + O\right)\left(\frac{1}{\sqrt{N}}\right),$$

the second equation by partial integration when evaluating the expectation for the $J_{ij}$-variables, and the last by Proposition 7.3. We now wish to argue that $\mathbb{E}[m_1^2 m_2^2] \approx q^2$:

**Proposition 7.7.** *There exists $\beta_o > 0$ such that for $\beta \leq \beta_o$ and all $h$*

$$\lim_{N \to \infty} \left|\mathbb{E}\left[m_1^2 m_2^2\right] - q^2(\beta, h)\right| = 0.$$

*(Remember that the $m_i$ are random variables, depending on $\beta, h, N$.)*

We postpone the proof of this for the moment, and finish the proof of Theorem 7.2: An elementary calculation yields

$$\frac{\partial}{\partial \beta}\left(\frac{\beta^2}{4}(1-q)^2 + \frac{1}{\sqrt{2\pi}}\int \log \cosh(h + \beta\sqrt{q}x)\exp\left[-\frac{x^2}{2}\right]dx\right) = \frac{\beta}{2}(1 - q^2(\beta, h)).$$

On the other hand, we get from Proposition 7.7

$$\lim_{N \to \infty} \int_0^{\beta_o} \left|\mathbb{E}\left(\frac{1}{N}\frac{\partial F_N}{\partial \beta}(\beta, h)\right) - \frac{\beta}{2}(1 - q^2(\beta, h))\right|d\beta = 0.$$

Using the (easy)

**Exercise 7.8.** $\lim_{N \to \infty} \frac{1}{N}F_N(0, \beta) = \log \cosh(h)$

we get

$$\lim_{N \to \infty} \mathbb{E}\left(\frac{1}{N}F_N(\beta, h)\right) = \log \cosh(h) + \int_0^{\beta} \frac{\beta}{2}(1 - q^2(\beta', h))\, d\beta'$$

$$= \frac{\beta^2}{4}(1-q)^2 + \frac{1}{\sqrt{2\pi}}\int \log \cosh(h + \beta\sqrt{q}x)\exp\left[-\frac{x^2}{2}\right]dx$$

for all $\beta \leq \beta_o$. Using again the concentration inequality of the last lecture (Proposition 6.7), we obtain in the same way as there the statement of the theorem.

*Proof of Proposition 7.7* . We define

$$\psi(x) = \frac{1}{\sqrt{2\pi}} \int \exp[-y^2/2] \tanh^2(xy+h) dy,$$

($h$ is fixed).

We now consider the system with $N+2$ spins, and at temperature parameter $\beta\sqrt{1+2/N}$. Expressions calculated in this system will get an upper index $(N+2)$, and we will indicate the temperature parameter in the notation, by writing for instance $m_{N+1}^{(N+2)}(\beta\sqrt{1+2/N})$ for the expectation of $\sigma_{N+1}$ calculated in the Gibbs measure of size $N+2$ and temperature parameter $\beta\sqrt{1+2/N}$. We apply the Remark 7.6 to $P = P^{(N)}$ (with the $J_{ij}$, $1 \le i < j \le N$ fixed for the moment), and $\xi_j^{(1)} = J_{j,N+1}$, $\xi_j^{(2)} = J_{j,N+2}$, and temperature parameter $\beta$. Remark now, that

$$m_{N+1}^{(N+2)}(\beta\sqrt{1+2/N}) = \frac{E^{(N)}[\sinh(Y_N^{(1)})\cosh(Y_N^{(2)})]}{E^{(N)}[\cosh(Y_N^{(1)})\cosh(Y_N^{(2)})]} + R_1$$

$$= \tanh\left(h + \frac{\beta}{\sqrt{N}} \sum_{j=1}^{N} J_{j,N+1} m_j^{(N)}(\beta)\right) + R_1 + R_2,$$

where $R_1$ is stemming from neglecting the direct $J_{N+1,N+2}/\sqrt{N}$ interaction, and $R_2$ is coming from the error term in Remark 7.6. The same expression holds for $m_{N+2}^{(N+2)}$ with $Y_N^{(1)}$ and $Y_N^{(2)}$ interchanged. Integrating first out the cavity variables $\xi_j^{(1)} = J_{j,N+1}$, $\xi_j^{(2)} = J_{j,N+2}$, $1 \le j \le N$, and then the other variables $J_{ij}$, $1 \le i < j \le N$ as well, one gets

$$\mathbb{E}(m_{N+1}^{(N+2)}(\beta\sqrt{1+2/N})^2 m_{N+2}^{(N+2)}(\beta\sqrt{1+2/N})^2)$$

$$= \mathbb{E}\left(\left[\psi\left(\frac{\beta}{\sqrt{N}}\sqrt{\sum_{j=1}^{N} m_j^{(N)^2}}\right)\right]^2\right) + o(1),$$

and analogously

$$\mathbb{E}(m_{N+2}^{(N+2)}(\beta\sqrt{1+2/N})^2) = \mathbb{E}\left(\psi\left(\frac{\beta}{\sqrt{N}}\sqrt{\sum_{j=1}^{N} m_j^{(N)^2}}\right)\right) + o(1).$$

These two equations yield

$$\mathrm{var}\left[\frac{1}{N+2}\sum_{j=1}^{N+2} m_j^{(N+2)}(\beta\sqrt{1+2/N})^2\right] = \mathrm{var}\left[\psi\left(\frac{\beta}{\sqrt{N}}\sqrt{\sum_{j=1}^{N} m_j^{(N)^2}}\right)\right] + o(1)$$

$$\le C\beta^2 \mathrm{var}\left(\frac{1}{N}\sum_{j=1}^{N} m_j^{(N)}(\beta)^2\right) + o(1),$$

as $x \to \psi(\sqrt{x})$ is easily seen to be Lipshitz continuous. This implies that for small enough $\beta > 0$

(7.8)
$$\lim_{N\to\infty} \mathrm{var}\left(\frac{1}{N}\sum_{j=1}^{N} m_j^{(N)2}\right) = 0.$$

On the other hand, using Proposition 7.5, we see that

$$m_{N+1}^{(N+1)}(\beta\sqrt{1+1/N}) = \frac{E^{(N)}\sinh(h+Y_N)}{E^{(N)}\cosh(h+Y_N)}$$

$$= \tanh\left(h + (\beta/\sqrt{N})\sum_{j=1}^{N} m_j^{(N)}(\beta)\, J_{j,N+1}\right) + R.$$

From this we get

(7.9)    $$\mathbb{E}\left[m_{N+1}^{(N+1)2}(\beta\sqrt{1+1/N})\right] = \mathbb{E}_N\left[\psi\left(\beta\sqrt{\frac{1}{N}\sum_{i=1}^{N} m_i^{(N)}(\beta)^2}\right)\right] + o(1)$$

Using (7.8), we get with $q_N(\beta) \stackrel{\text{def}}{=} \mathbb{E}m_1^{(N+1)2}(\beta\sqrt{1+1/N})$

$$q_{N+1}\left(\beta\sqrt{1+\frac{1}{N}},h\right) = \psi(\beta\sqrt{q_N(\beta,h)},h) + o(1).$$

From this relation, one can easily derive, that for small enough $\beta$

$$\lim_{N\to\infty} q_N(\beta,h) = q(\beta,h).$$

Together with (7.8), this proves the Proposition.    □

# Lecture 8:

# The Random Energy Model

### 8.1. The Free Energy of the Random Energy Model

In an attempt to gain some understanding of the low temperature regime in mean-field spin glasses, Derrida [16] investigated the so-called Random Energy Model (see also [45]). Remark that in the SK-model (with $h = 0$, say), the Hamiltonian

$$H(\sigma) = \frac{1}{\sqrt{N}} \sum_{1 \leq i < j \leq N} J_{i,j} \sigma_i \sigma_j$$

is a family of random variables, indexed by the $\sigma \in \Sigma_N$, which are normally distributed with variance $(N - 1)/2$. The main difficulty in the SK-model is hidden in the fact that these random variables are not independent. In fact, $\mathbb{E} H(\sigma) H(\sigma') = (1/2N)(\langle \sigma, \sigma' \rangle^2 - N)$. One may however ask whether some interesting features show up when assuming that these variables are just independent Gaussian random variables. There is no point to stick to the variance $(N - 1)/2$, and we take the variance $N$. Evidently then, the $\sigma$ need not carry any internal structure. We therefore assume that we have just $2^N$ independent Gaussian random variables, say $X_\alpha^{(N)}$, $1 \leq \alpha \leq 2^N$, defined on some probability space $(\Omega, \mathcal{F}, \mathbb{P})$, which are centered and have variance $N$. We define the "Gibbs measure" on the set $\Sigma_N$ of "configurations" $\alpha$'s by defining for any $\omega \in \Omega$, and any $\beta > 0$

$$P_{\omega,\beta,N}(\alpha) = 2^{-N} \frac{\exp[\beta X_\alpha^{(N)}(\omega)]}{Z_{\omega,\beta,N}},$$

where $Z_{\omega,\beta,N} = \sum_a 2^{-N} \exp[\beta X_\alpha^{(N)}(\omega)] = E_o e^{\beta X.(\omega)}$, $P_o$ being again the uniform distribution on $\Sigma_N$. For any fixed $\omega \in \Omega$, this is a probability distribution on the $\alpha$, and we would like to know how it behaves for $N \to \infty$, and almost all $\omega$. The first task is to investigate the free energy

$$f(\beta) = \lim_{N \to \infty} \frac{1}{N} \log Z_{\omega,\beta,N}.$$

This could still depend on $\omega$, but we will see in a moment that it does not. In fact, we have the following result:

**Theorem 8.1.** $f(\beta)$ exists almost surely and is given by

$$f(\beta) = \begin{cases} \dfrac{\beta^2}{2} & \text{if } \beta \leq \sqrt{2 \log 2} \\ \sqrt{2 \log 2} \beta - \log 2 & \text{if } \beta \geq \sqrt{2 \log 2} \end{cases}.$$

The high temperature value is again the annealed free energy

$$\lim_{N \to \infty} \frac{1}{N} \log \mathbb{E} Z_{\beta,N}.$$

Curiously, it is *not* true that

$$(8.1) \qquad \qquad \mathbb{E} Z_\beta^2 \leq C \left( \mathbb{E} Z_\beta \right)^2$$

up to the correct critical value $\beta_c = \sqrt{2\log 2}$, which contrasts with the situation in the SK-model at $h = 0$:

$$\mathbb{E}Z_\beta^2 = \sum_{\alpha,\alpha'} 2^{-2N} \mathbb{E}\exp[\beta(X_\alpha^{(N)} + X_{\alpha'}^{(N)})]$$

$$= \sum_\alpha 2^{-2N}\exp[2\beta^2 N] + \sum_{\alpha\neq\alpha'} 2^{-2N}\exp[\beta^2 N]$$

$$= \exp[N(-\log 2 + 2\beta^2)] + \{1 - 2^{-N}\}\exp[N\beta^2].$$

The first summand dominates the second as soon as $\beta > \sqrt{\log 2}$, and in fact, $\mathbb{E}Z_\beta^2/(\mathbb{E}Z_\beta)^2$ is exponentially growing in this case.

**Proof of Theorem 8.1.** The trick is to apply the "second moment method" not directly to $Z$ but to

$$A_N(s) \overset{\text{def}}{=} \#\{\alpha : X_\alpha^{(N)} \geq sN\}.$$

Let $\Phi$ be the standard normal distribution function. Then

$$\mathbb{E}A_N(s) = 2^N(1 - \Phi(s\sqrt{N})) \asymp 2^N e^{-s^2 N/2},$$

Remark that for $s > \sqrt{2\log 2}$, $\mathbb{E}A_N(s)$ converges to 0, exponentially in $N$. Using the Markov inequality and the Borel-Cantelli Lemma, we can conclude that

$$(8.2) \qquad\qquad A_N(s) = 0, \qquad \mathbb{P}\text{-a.s.}$$

for large enough $N$. If we calculate now

$$(8.3) \qquad \mathbb{E}A_N(s)^2 = 2^N(1 - \Phi(s\sqrt{N})) + 2^N(2^N - 1)(1 - \Phi(s\sqrt{N}))^2,$$

we see that for $s < \sqrt{2\log 2}$, this is $[\mathbb{E}A_N(s)]^2$, up to a factor, which is exponentially close to 1. Using the Borel-Cantelli Lemma, we get that for any $s_o < \sqrt{2\log 2}$ and all $\varepsilon > 0$, one has with probability one

$$(8.4) \qquad \exp\left[N\left(\log 2 - \frac{s^2}{2} - \varepsilon\right)\right] \leq A_N(s) \leq \exp\left[N\left(\log 2 - \frac{s^2}{2} + \varepsilon\right)\right]$$

for all $s \in [0, s_o]$ and large enough $N$. On the other hand, one has for any $\delta > 0$

$$2^{-N}\sum_\alpha e^{\beta X_\alpha^{(N)}} = N\beta 2^{-N}\int_{-\infty}^\infty A_N(s)e^{N\beta s}ds$$

$$\leq N\beta 2^{-N}\int_{-\infty}^{\sqrt{2\log 2}+\delta} A_N(s \wedge (\sqrt{2\log 2} - \delta))e^{N\beta s}\, ds$$

and

$$\geq N\beta 2^{-N}\int_{-\infty}^{\sqrt{2\log 2}-\delta} A_N(s)e^{N\beta s}\, ds,$$

$\mathbb{P}$-a.s. for large enough $N$, where we used (8.2). Applying (8.4) and letting first $N \to \infty$ and afterwards $\varepsilon, \delta \to 0$, we get

$$\lim_{N \to \infty} \frac{1}{N} \log 2^{-N} \sum_{\alpha} e^{\beta X_{\alpha}^{(N)}} = \sup_{s \leq \sqrt{2 \log 2}} \left\{ -\frac{s^2}{2} + \beta s \right\}$$

$$= \begin{cases} \dfrac{\beta^2}{2} & \text{if } \beta \leq \sqrt{2 \log 2} \\ \sqrt{2 \log 2} \beta - \log 2 & \text{if } \beta \geq \sqrt{2 \log 2} \end{cases}.$$

This finishes the proof of Theorem 8.1.

We next want to describe the Gibbs measure $P_{\omega,\beta,N}$ in the $N \to \infty$ limit. We have to distinguish between the high temperature case $\beta < \sqrt{2 \log 2}$, and the low temperature case $\beta > \sqrt{2 \log 2}$. We abstain from discussing the critical case $\beta = \sqrt{2 \log 2}$. The fundamental difference is that in the high temperature case, the Gibbs measure is concentrated on a growing number of energy levels, which get denser and denser packed as $N \to \infty$. In contrast, in the low temperature regime, the Gibbs distribution is essentially concentrated on the top energy levels as will be discussed in Section 8.3.

Some rough information on the high temperature case is easy, and is collected in the following result:

**Proposition 8.2.** *Assume $\beta < \sqrt{2 \log 2}$.*

a) *Given $\varepsilon > 0$ there exists $\delta > 0$ such that*

$$\mathbb{P}(P_{\beta,N}(\{ \alpha : (\beta - \varepsilon)N \leq X_{\alpha}^{(N)} \leq (\beta + \varepsilon)N \}) \geq 1 - e^{-\delta N}) \geq 1 - e^{-\delta N}.$$

b) *For some $\delta = \delta(\beta) > 0$*

$$\lim_{N \to \infty} e^{N\delta} \sup_{\alpha} P_{\beta,N}(\alpha) = 0 \quad \mathbb{P}\text{-}a.s.$$

*Proof.* Remark first that for $\beta < \sqrt{2 \log 2}$, $\sup_{s \leq \sqrt{2 \log 2}}(-\frac{s^2}{2} + \beta s)$ is uniquely attained at $s = \beta$ and therefore, for any $\varepsilon > 0$

$$\alpha(\beta, \varepsilon) \stackrel{\text{def}}{=} \sup_{s \notin [\beta - \varepsilon, \beta + \varepsilon]} \left( -\frac{s^2}{2} + \beta s \right) < \frac{\beta^2}{2}.$$

Using the same argument as in the proof of Theorem 8.1, we get

$$\limsup_{N \to \infty} \frac{1}{N} \log 2^{-N} \sum_{\alpha : \frac{X_{\alpha}^{(N)}}{N} \notin [\beta - \varepsilon, \beta + \varepsilon]} e^{\beta X_{\alpha}^{(N)}} \leq a(\beta, \varepsilon) \quad \mathbb{P}\text{-}a.s.$$

Using the variance estimate from (8.3), and the Tchebychev inequality, it is easily seen that

$$\mathbb{P}(\Lambda_N^c(\beta,\varepsilon)) \leq \exp(-\delta' N)$$

for some $\delta' > 0$, where $\Lambda_N(\beta,\varepsilon)$ is the event

$$\left\{2^{-N} \sum_{\alpha:\frac{X_\alpha^{(N)}}{N} \notin [\beta-\varepsilon,\beta+\varepsilon]} e^{\beta X_\alpha^{(N)}} \leq \alpha(\beta,\varepsilon) + \frac{\eta}{3}, \ 2^{-N} \sum_\alpha e^{\beta X_\alpha^{(N)}} \geq \alpha(\beta,\varepsilon) + \frac{2\eta}{3}\right\},$$

$$\eta \overset{\text{def}}{=} \frac{\beta^2}{2} - \alpha(\beta,\varepsilon).$$

This proves a) with $\delta \overset{\text{def}}{=} \delta' \wedge (\eta/3)$.

To prove b), first observe that a) implies that

$$\max_{\alpha:X_\alpha^{(N)}>(\beta+\varepsilon)N} P_{\beta,N}(\alpha) \leq e^{-\delta n}$$

with $\mathbb{P}$-probability $\geq 1 - e^{-\delta N}$. We chop the interval $[0,(\beta+\varepsilon)N]$ into intervals of length 1. So we have $(\beta+\varepsilon)N$ such intervals, $[j, j+1)$. The expected number $\Xi_j$ of "configurations" $\alpha$ with $X_\alpha^{(N)} \in [j, j+1)$ is still exponential in $N$ (if $\beta+\varepsilon < \sqrt{2\log 2}$), and a similar computation as in (8.3) gives a variance

$$\text{var}_{\mathbb{P}}(\Xi_j) \leq e^{-\delta N} \mathbb{E}(\Xi_j)^2$$

for some $\delta > 0$ (and large enough $N$). We therefore conclude that

$$\mathbb{P}(\Xi_j \geq e^{\delta N}, \forall j) \geq 1 - e^{-\delta N}$$

again for some $\delta > 0$. However, if $\Xi_j \geq e^{\delta N}$, then for any $\alpha$ with $X_\alpha^{(N)} \in [j, j+1)$ have

$$P_{\beta,N}(\alpha) \leq e^{\beta(j+1)}/e^{\beta j}\Xi_j \leq e^\beta e^{-\delta N}.$$

Using again Borel-Cantelli, we see that with $\mathbb{P}$-probability one

$$\lim_{N\to\infty} e^{\delta N} \sup_\alpha P_{\beta,N}(\alpha) = 0$$

for some $\delta \ (= \delta(\beta))$.                                                    □

The low temperature case (i.e. $\beta$ large) is much more interesting. In order to treat it, we need some facts from point-process theory.

## 8.2. A Short Introduction to Point Processes

We collect the results which we need in the next section. For proofs, the reader is referred to [53].

Let $I$ be an open interval in $\mathbb{R}$, possibly $\mathbb{R}$ itself. We consider Radon measures in $(I, \mathcal{B}_I)$ : These are measures $\mu$ which have the property that $\mu(K) < \infty$ for any compact subset of $I$. We denote the set of Radon measures by $M(I)$.

$M(I)$ is naturally equipped with the $\sigma$–field $\mathcal{M}(I)$, which is generated by by the evaluation mapping $\mu \to \mu(A)$, $A \in \mathcal{B}_I$. The topology generated by the mappings $M(I) \ni \mu \to \int f d\mu$, $f \in C_o(I)$ is called the *vague topology*. $C_o(I)$ is the set of continuous real-valued functions of compact support. This topology is metrizable with a separable complete metric, i.e. $M(I)$ is a Polish space. $\mathcal{M}(I)$ is the Borel field for this topology. Point measures are special elements in $M(I)$ of the form

$$(8.5) \qquad\qquad\qquad \mu = \sum_n \delta_{x_n},$$

where $(x_n)$ is a countable sequence of points with the property that $\#\{n : x_n \in K\} < \infty$ for any compact subset $K$ of $I$. We denote by $M_p(I)$ the set of such measures. It is readily checked that $M_p(I) \in \mathcal{M}(I)$. In fact $M_p(I)$ is a closed subset of $M(I)$. We denote by $\mathcal{M}_p(I)$ the trace of $\mathcal{M}(I)$ on $M_p(I)$. Measures in $M_p(I)$ may of course charge single points with natural numbers other than 1, but the ones we consider will not do that. We call such measures *pure point measures*, which can be written as (8.5) with all $x_n$ distinct. We often will encounter the situation where the set $\{x_n : n \in \mathbb{N}\}$ has a maximal element. We can then order it downwards: $x_0 > x_1 > x_2 > \ldots$, and this sequence has no accumulation point in $I$. We write $M_>(I)$ for the set of pure point measures obtained in this way. $M_>(I)$ is neither open nor closed in $M(I)$, but it is readily checked that it is a Borel subset. Therefore we can again take the trace of $\mathcal{M}(I)$ on this set, denoted by $\mathcal{M}_>(I)$.

A point process is a random variable taking values in $(M_p(I), \mathcal{M}_p(I))$. Its distribution is then a probability measure on this measurable space. For probability measures on $(M_p(I), \mathcal{M}_p(I))$ we have the notion of weak convergence, meaning convergence of the integrals over all bounded continuous functions $M_p(I) \to \mathbb{R}$.

If a point process takes values in $M_>(I)$, we often write it is a sequence $(\eta_n)_{n \geq 0}$ of real valued random variables which are order downwards. We will sometimes be a bit careless and also talk of the point process $(X_a)_{a \in A}$ if this is a countable set of random variables which with probability one are disjoint, have no finite accumulation point, and have a maximal element. Of course, we then mean the random variable $\sum_a \delta_{X_a}$, taking values in $(M_>, \mathcal{M}_>)$. We write $\overset{w}{\to}$ for weak convergence. By a slight abuse of notation we occasionally write $(\eta_n^{(N)})_{n \geq 0} \overset{w}{\to} Q$, $N \to \infty$, where $Q$ is a probability measure on $(M_p(I), \mathcal{M}_p(I))$, meaning, of course, that the laws of the $(\eta_n^{(N)})_{n \geq 0}$ converge as $N \to \infty$.

Of special importance are Poisson point processes. Let $F$ be a Radon measure on $I$. The Poisson point process with intensity $F$ is a point process whose distribution $Q_F$ on $(M_p(I), \mathcal{M}_p(I))$ is characterized by the following properties:

- For any compact subset $A \subset I$, $M_p(I) \ni \mu \to \mu(A)$ is under $Q_F$ Poisson distributed with parameter $F(A)$.
- If $A_1, \ldots, A_k$ are disjoint compact subsets of $I$, then the variables $\mu(A_1)$, $\ldots, \mu(A_k)$ are independent.

It is a standard fact in point process theory that for any Radon measure $F$ such a measure exists and is uniquely characterized by these properties. We call $Q_F$ a Poisson measure. In all cases we are interested in, $F$ will have a density $f$. $Q_F$ is concentrated on pure point measures if $F$ has a density. By an abuse of notation we also call $f$ the intensity. We will write $PPP(f)$ or $PPP(t \to f(t))$ for the Poisson point process with this intensity (strictly speaking its law).

A convenient tool for the investigation of point processes are Laplace functionals: Let $\phi \in C_o^+(I)$, and $Q$ be a probability measure on $(M(I), \mathcal{M}(I))$. Then

$$L_\phi(Q) \stackrel{\text{def}}{=} \int \exp\left[-\int \phi d\mu\right] dQ.$$

If $L_\phi(Q) = L_\phi(Q')$ for all $\phi \in C_0^+(I)$ then $Q = Q'$.

**Exercise 8.3.** *If $Q_F$ is Poissonian with Radon measure $F$, then*

$$L_\phi(Q_F) = \exp\left[-\int \left(1 - e^{-\phi(x)}\right) F(dx)\right].$$

**Exercise 8.4.** *$\{Q_n\}_{n \in \mathbf{N}}$ converges weakly to $Q$ if and only if for any $\phi \in C_o^+(I)$ :*

$$\lim_{n \to \infty} L_\phi(Q_n) = L_\phi(Q).$$

An important property of Poisson point processes is that they transform nicely under mappings. We need that only in a special case.

Let $\psi : I \to I'$ be a continuous mapping having the property that $\psi^{-1}(K)$ is compact in $I$ whenever $K$ is compact in $I'$. Then $\psi$ defines a mapping $M(I) \to M(I')$, $\mu \to \mu\psi^{-1}$, denoted by $\Psi$.

**Proposition 8.5.** *Le $F$ be a Radon measure on $I$ and $Q_F$ be the corresponding Poisson measure. Then $Q_F \Psi^{-1} = Q_{F\psi^{-1}}$.*

*Proof.* If $\phi \in C_0^+(I')$ then

$$L_\phi(Q_F \Psi^{-1}) = \int \exp\left(-\int \phi \, d\mu\right) Q_F \Psi^{-1}(d\mu)$$

$$= \int \exp\left(-\int (\phi \circ \psi) \, d\mu\right) Q_F(d\mu)$$

$$= \exp\left[-\int (1 - e^{-\phi \circ \psi}) \, dF\right]$$

$$= \exp\left[-\int (1 - e^{-\phi}) d(F\psi^{-1})\right].$$

$\square$

### 8.3. The Limiting Behavior of the Random Energy Model

We apply this now to our Random Energy Model. Remember that the random variables $X_\alpha^{(N)}$, $1 \leq \alpha \leq 2^N$ had been normally distributed with mean 0 and variance $N$. Remark that the $X_\alpha^{(N)}$ are all different, with probability one. Therefore, for any sequence $a_N$ of real numbers, $(X_\alpha^{(N)} - a_N)_{\alpha \in \Sigma_N}$ defines a pure point process on $\mathbb{R}$. This is a slight abuse of notation. Strictly speaking $\sum_\alpha \delta_{X_\alpha^{(N)} - a_N}$ is the point process, but we usually keep the simpler notation.

**Proposition 8.6.** *If*

$$a_N = \sqrt{2 \log 2} N - \frac{1}{2\sqrt{2 \log 2}} \log N - \frac{\log 2 + \frac{1}{2}(\pi \log 2)}{\sqrt{2 \log 2}},$$

*then the above point process converges weakly to* $PPP(t \to \sqrt{2 \log 2} e^{-\sqrt{2 \log 2} t})$.

*Proof.* We denote by $Q_N$ the law of $\sum_\alpha \delta_{X_a - a_N}$ . If $\phi \in C_o^+(I)$, the

$$
\begin{aligned}
L_\phi(Q_N) &= \mathbb{E} \exp\left[ - \sum_\alpha \phi(X_\alpha^{(N)} - a_N) \right] \\
&= \left\{ \frac{1}{\sqrt{2\pi N}} \int \exp\left[ -\phi(x - a_N) - \frac{x^2}{2N} \right] dx \right\}^{2^N} \\
&= \left\{ 1 - \frac{1}{\sqrt{2\pi N}} \int (1 - e^{-\phi(x)}) \exp\left[ -\frac{(x + a_N)^2}{2N} \right] dx \right\}^{2^N} \\
&= \exp\left[ -2^N \frac{1}{\sqrt{2\pi N}} \int (1 - e^{-\phi(x)}) \exp\left[ -\frac{(x + a_N)^2}{2N} \right] dx \right] (1 + o(1)).
\end{aligned}
$$

Furthermore

$$\exp\left[ -\frac{(x + a_N)^2}{2N} \right] = \sqrt{4\pi \log 2} e^{-x\sqrt{2 \log 2}} \exp\left[ -N \log 2 \right] \sqrt{N} (1 + o(1)),$$

in the domain of integration. Therefore

$$\lim_{N \to \infty} L_\phi(Q_N) = \exp\left[ -\sqrt{2 \log 2} \int (1 - e^{-\phi(x)}) \exp[-\sqrt{2 \log 2} x] dx \right],$$

as required.                                                                  □

The Poisson point processes with intensity $t \to a e^{-at}$ have a number of remarkable properties. Let $(\eta_i)$ such a point process described by the random points $\eta_0 > \eta_1 > \dots$.

**Proposition 8.7.**

a) *Let $X_i, i \in \mathbb{N}$, be a sequence of i.i.d. random variables satisfying $M(a) \stackrel{\text{def}}{=} \mathbb{E}(e^{aX_i}) < \infty$, which is also independent of $(\eta_i)$. Then*

$$\mathcal{L}\left( \left( \eta_i + X_i - \frac{1}{a} \log M(a) \right)_{i \in \mathbb{N}} \right) = \mathcal{L}((\eta_i)_{i \in \mathbb{N}}),$$

*where $\mathcal{L}(\cdot)$ denotes the law.*

b) *Let* $(\eta_i^k)$, $k \in \mathbb{N}$ *be an i.i.d. sequence of* $PPP(t \to ae^{-at})$, *and let* $x_k$, $k \in \mathbb{N}$, *be a sequence of real numbers with* $M(a) \stackrel{def}{=} \sum_k e^{ax_k} < \infty$. *Then*

$$\mathcal{L}\left(\left(\eta_i^k + x_k - \frac{1}{a} \log M(a)\right)_{i,k \in \mathbb{N}}\right) = \mathcal{L}((\eta_i)_{i \in \mathbb{N}}).$$

*Proof.* We prove a). b) is similar. Let $\phi \in C_o^+(\mathbb{R})$. Then

$$\mathbb{E}\left(\exp\left[-\sum_i \phi(\eta_i + X_i)\right]\right) = \mathbb{E}_\eta \prod_i \mathbb{E}_X \exp\left[-\phi(\eta_i + X)\right]$$
$$= \mathbb{E}_\eta \prod_i \exp\left[-\psi(\eta_i)\right],$$

where $e^{-\psi(x)} = \int e^{-\phi(x+y)} F(dy)$, $F$ being the distribution of the $X_i$. $\mathbb{E}_\eta$ denotes taking expectation with respect to the point process and $\mathbb{E}_X$ with respect to the $X$-variables. We then get

$$\mathbb{E}\left(\exp\left[-\sum_i \phi(\eta_i + X_i)\right]\right) = \exp\left[-a\int(1 - e^{-\psi(x)})e^{-ax}dx\right]$$
$$= \exp\left[-a\int\int(1 - e^{-\phi(x+y)})e^{-ax}dx\, F(dy)\right]$$
$$= \exp\left[-a\int(1 - e^{-\phi(x)})e^{-ax}dx \int e^{ay} F(dy)\right]$$
$$= \exp\left[-a\int(1 - e^{-\phi(x)})\exp\left[-a\left(x - \frac{1}{a}\log M(a)\right)\right]dx\right].$$

This proves the claim. b) is similar.                                   □

We now discuss the limiting Gibbs distribution of the Random Energy Model for $\beta > \sqrt{2 \log 2}$. First remark that applying Proposition 8.5 to the function $\mathbb{R} \ni y \to \exp[\beta y] \in (0, \infty)$, we obtain:

**Corollary 8.8.** *The point process*

$$(\exp(\beta(X_\alpha^{(N)} - a_N)))_{\alpha \in \Sigma_N}$$

*converges weakly as* $N \to \infty$ *to* $PPP(t \to xt^{-x-1})$, *where*

$$x = x(\beta) = \frac{\sqrt{2 \log 2}}{\beta} \in (0, 1).$$

The Poisson point processes with this intensity play an absolutely crucial role in the Parisi theory of spin glasses. As the intensity is integrable at $\infty$, the point configuration have a maximal element, and we can interprete the point process as being a probability distribution on decreasing sequences $(\eta_l)_{l \in \mathbb{N}}$ of positive real numbers. One of the basic properties is

**Exercise 8.9.** *If* $(\eta_l)$ *is a* $PPP(t \to xt^{-x-1})$ *with* $0 < x < 1$ *then* $\sum_l \eta_l < \infty$ *almost surely.*

Using this fact, we can transform the point process by normalizing the $\eta_l$ :

$$\overline{\eta}_l \stackrel{\text{def}}{=} \frac{\eta_l}{\sum_i \eta_i}.$$

Evidently, this defines a point process $(\overline{\eta}_l)_{l \in \mathbb{N}}$, which is living on positive point configurations which sum up to 1. Such a point process cannot be Poissonian. We write $\mathcal{N}((\eta_l)) \stackrel{\text{def}}{=} (\overline{\eta}_l)$. $\mathcal{N}$ is not defined on all measures on $M_p(\mathbb{R}^+)$, and is also not quite continuous.

Coming now back to our Gibbs distribution, we see that the distribution is no changed by subtracting from the energy levels some constant, for instance the $a_N$ we had encountered before:

$$P_{\omega,\beta,N}(\alpha) = \frac{\exp[\beta X_\alpha^{(N)}]}{\sum_\alpha \exp[\beta X_\alpha^{(N)}]} = \frac{\exp[\beta(X_\alpha^{(N)} - a_N)]}{\sum_\alpha \exp[\beta(X_\alpha^{(N)} - a_N)]}.$$

Of course, we may regard this Gibbs measure as a point process $(P_{\cdot,\beta,N}(\alpha))_{\alpha \in \Sigma_N}$ which is living on point configurations which sum up to 1. On the background of the above discussion, it is now quite plausible that the following result is true

**Theorem 8.10.** *The point process $\{P_{\cdot,\beta,N}(\alpha)\}_{\alpha \in \Sigma_N}$ converges weakly to $\mathcal{N}((\eta_l))$, where $(\eta_l)$ is a Poisson point process with intensity $t \to xt^{-x-1}$, where $x = \sqrt{2\log 2}/\beta$.*

*Proof.* $\mathcal{N}$ is not quite a continuous transformation. However, one only has to check that

$$\lim_{K \to \infty} \sup_N \mathbb{P}\left( \sum_\alpha (1_{\xi_\alpha \leq 1/K} + 1_{\xi_\alpha \geq K}) \xi_\alpha \geq \delta \right) = 0$$

for any $\delta > 0$, where $\xi_\alpha = \exp(\beta(X_\alpha^{(N)} - a_N))$. $\qquad \square$

I leave this as an exercise.

# Lecture 9:

# The Generalized Random Energy Model and Induced Clusterings

### 9.1. The Generalized Random Energy Model (GREM)

This model has an in-built hierarchical structure with a finite number of hierarchical levels (see [24]). Later, we will generalize some aspects of the model to a "continuum" of levels. This is of crucial importance for the Parisi picture of the low temperature regime of the SK-model. We start with discussing the model with two levels. The model with finitely many levels will be a straightforward generalization.

We again would like to have $2^N$ energy levels $X_\alpha$, each normally distributed with variance $N$. We equip them however with nonvanishing correlations in the following way: We consider $\alpha$ to be of the form $\alpha = (\alpha_1, \alpha_2)$, where $1 \leq \alpha_i \leq 2^{N/2}$. (For simplicity, we assume that $N$ is even). We then take independent centered normally distributed random variables $X^1_{\alpha_1}$, $X^2_{\alpha_1,\alpha_2}$, where the $X^1$ have variance $\sigma_1^2 N$ and the $X^2$ have variance $\sigma_2^2 N$. We assume that $\sigma_1^2 + \sigma_2^2 = 1$, and define

$$X_\alpha = X^1_{\alpha_1} + X^2_{\alpha_1,\alpha_2}.$$

Evidently, the $X_\alpha$ are normally distributed with variance $N$. Remark that $\mathbb{E}(X_\alpha X_{\alpha'}) = 0$ if $\alpha_1 \neq \alpha_1'$, and $\mathbb{E}(X_\alpha X_{\alpha'}) = \sigma_1^2 N$ if $\alpha_1 = \alpha_1'$, $\alpha_2 \neq \alpha_2'$. We make the same analysis as in the REM case and determine first the free energy, and then the limiting Gibbs distribution.

Let's start with the free energy:

$$f(\beta) \stackrel{\text{def}}{=} \lim_{N \to \infty} \frac{1}{N} \log 2^{-N} \sum_\alpha e^{\beta X_\alpha}.$$

We again try some variant of the second moment method, but there is a surprise: If we consider $A_N(s) \stackrel{\text{def}}{=} \#\{\alpha : X_\alpha \geq sN\}$ as in the case of the Random Energy Model, then $\mathbb{E}A_N^2(s) \gg (\mathbb{E}A_N(s))^2$, for values of $s$ which are too small to get the correct picture. The reader is asked to check that. We therefore have to rely on a slightly more subtle procedure: We consider instead the random variables

$$A_N(s_1, s_2) \stackrel{\text{def}}{=} \#\{\alpha : X^1_{\alpha_1} \geq s_1 N, \ X^2_{\alpha_1,\alpha_2} \geq s_2 N\}.$$

Evidently

$$\mathbb{E}A_N(s_1, s_2) \asymp 2^N \exp\left[-\frac{s_1^2 N}{2\sigma_1^2} - \frac{s_2^2 N}{2\sigma_2^2}\right],$$

and

$$
\begin{aligned}
\mathbb{E}A_N^2(s_1, s_2) &= \sum_{\alpha, \alpha'} \mathbb{P}\big(X_{\alpha_1}^1 \geq s_1 N,\, X_{\alpha_1'}^1 \geq s_1 N,\, X_{\alpha_1, \alpha_2}^2 \geq s_2 N,\, X_{\alpha_1', \alpha_2'}^2 \geq s_2 N\big) \\
&= \sum_{\alpha_1 \neq \alpha_1'} \mathbb{P}(X^1 \geq s_1 N)^2\, \mathbb{P}(X^2 \geq s_2 N)^2 \\
&\quad + \sum_{\alpha_1 = \alpha_1', \alpha_2 \neq \alpha_2'} \mathbb{P}(X^1 \geq s_1 N)\mathbb{P}(X^2 \geq s_2 N)^2 + \sum_\alpha \mathbb{P}(X^1 \geq s_1 N)\mathbb{P}(X^2 \geq s_2 N) \\
&\asymp (2^{2N} - 2^{3N/2}) \exp\left[ -\frac{s_1^2 N}{\sigma_1^2} - \frac{s_2^2 N}{\sigma_2^2} \right] \\
&\quad + (2^{3N/2} - 2^N) \exp\left[ -\frac{s_1^2 N}{2\sigma_1^2} - \frac{s_2^2 N}{\sigma_2^2} \right] + 2^N \exp\left[ -\frac{s_1^2 N}{2\sigma_1^2} - \frac{s_2^2 N}{2\sigma_2^2} \right].
\end{aligned}
$$

The question now is again under which conditions the "leading summand" above, namely $2^{2N} \exp[-s_1^2 N/\sigma_1^2 - s_2^2 N/\sigma_2^2]$ dominates $\mathbb{E}A_N^2(s_1, s_2)$, i.e. that

$$
\mathbb{E}A_N^2(s_1, s_2) - (\mathbb{E}A_N(s_1, s_2))^2 \ll (\mathbb{E}A_N(s_1, s_2))^2,
$$

which by Chebychev will imply that

$$
A_N(s_1, s_2) \simeq \mathbb{E}A_N(s_1, s_2)
$$

with large probability. A straightforward computation yields that this is true if the following two conditions are satisfied:

$$
\text{(9.1)} \qquad\qquad\qquad \frac{s_1^2}{\sigma_1^2} < \log 2
$$

$$
\text{(9.2)} \qquad\qquad\qquad \frac{s_1^2}{2\sigma_1^2} + \frac{s_2^2}{2\sigma_2^2} < \log 2.
$$

If these two conditions are met, then by Chebyshev and Borel-Cantelli, we have

$$
\left| \frac{A_N(s_1, s_2)}{\mathbb{E}A_N(s_1, s_2)} - 1 \right| \leq e^{-\delta(s_1, s_2)N}
$$

with some $\delta(s_1, s_2) > 0$, for large enough $N$, $\mathbb{P}$-a.s.. On the other hand, $A_N(s_1, s_2) = 0$ if $s_1^2/\sigma_1^2 > \log 2$ or $s_1^2/2\sigma_1^2 + s_2^2/2\sigma_2^2 > \log 2$, for large $N$, $\mathbb{P}$-a.s. Remark that for $\sigma_1^2 > \sigma_2^2$ we may have $s_1^2/\sigma_1^2 > \log 2$ and $s_1^2/2\sigma_1^2 + s_2^2/2\sigma_2^2 < \log 2$ for appropriate choices of $s_1$ and $s_2$, and therefore for such $s_1$, $s_2$, $A_N(s_1, s_2) = 0$ for large $N$, a.s., despite that $\mathbb{E}A_N(s_1, s_2) \to \infty$, exponentially fast. Using the same analysis as for the REM, we therefore get

$$
f(\beta) = \lim_{N \to \infty} \frac{1}{N} \log 2^{-N} \sum_\alpha e^{\beta X_\alpha}
$$

$$
\text{(9.3)} \qquad = \sup\left\{ \beta(s_1 + s_2) - \frac{s_1^2}{2\sigma_1^2} - \frac{s_2^2}{2\sigma_2^2} : \frac{s_1^2}{\sigma_1^2} \leq \log 2,\, \frac{s_1^2}{\sigma_1^2} + \frac{s_2^2}{\sigma_2^2} \leq 2\log 2 \right\}.
$$

Maximizing over $s_1, s_2$ with $s_1 + s_2 = t$ fixed and disregarding the side conditions

(9.4)
$$\frac{s_1^2}{\sigma_1^2} \leq \log 2,$$

(9.5)
$$\frac{s_1^2}{\sigma_1^2} + \frac{s_2^2}{\sigma_2^2} \leq 2\log 2$$

we have the maximum attained for

(9.6)
$$\frac{s_1}{\sigma_1^2} = \frac{s_2}{\sigma_2^2} = t.$$

We now have to distinguish two cases. Let us first look at the uninteresting one:

**Case I:** $\sigma_1^2 \leq \sigma_2^2$. In this case, we see that under the optimal choice (9.6), the condition (9.4) is implied by (9.5). Therefore, we get in this case

$$f(\beta) = \sup\left\{\beta t - \frac{t^2}{2} : t \leq \sqrt{2\log 2}\right\},$$

which is exactly the free energy of the REM. More interesting is

**Case II:** $\sigma_1^2 > \sigma_2^2$. In this case, the unrestricted maximum is at $s_1 = \beta\sigma_1^2$, $s_2 = \beta\sigma_2^2$, which is optimizing (9.3), as long as the first condition (9.4) is satisfied, as the second (9.5) is implied by the first. We therefore get $f(\beta) = \beta^2/2$ as long as

$$\beta \leq \beta_1^{\mathrm{cr}} \stackrel{\text{def}}{=} \frac{\sqrt{\log 2}}{\sigma_1}.$$

If this is violated, we have just have to take $s_1 = \sigma_1\sqrt{\log 2}$ in our variational expression (9.3) and optimize with respect to $s_2$. Summarizing, we get

**Proposition 9.1.** *Assume $\sigma_1^2 > \sigma_2^2$, then*

$$f(\beta) = \begin{cases} \beta^2/2 & \text{for} \quad 0 \leq \beta \leq \beta_1^{\mathrm{cr}} \\ \beta\sigma_1\sqrt{\log 2} - \dfrac{\log 2}{2} + \dfrac{\beta^2\sigma_2^2}{2} & \text{for} \quad \beta_1^{\mathrm{cr}} \leq \beta \leq \beta_2^{\mathrm{cr}} \\ \beta(\sigma_1 + \sigma_2)\sqrt{\log 2} - \log 2 & \text{for} \quad \beta_2^{\mathrm{cr}} \leq \beta \end{cases},$$

*where $\beta_i^{\mathrm{cr}} \stackrel{\text{def}}{=} \frac{\sqrt{\log 2}}{\sigma_i}$.*

We now discuss the Gibbs measures. The case $\sigma_1^2 \leq \sigma_2^2$ offers no surprise: The outcome is exactly the same as in the REM-case (with some slight complications for $\sigma_1^2 = \sigma_2^2$). For the rest of the discussion, we therefore assume $\sigma_1^2 > \sigma_2^2$. A rough picture of what happens is obtained from the above evaluation of the free energy. We have used there that $\max_{\alpha_1} X_{\alpha_1}^1 \approx N\sqrt{\log 2}\,\sigma_1$, and for any $\alpha_1$, $\max_{\alpha_2} X_{\alpha_1,\alpha_2}^2 \approx N\sqrt{\log 2}\,\sigma_2$. If $\beta < \beta_1^{\mathrm{cr}}$ then the main contribution to the partition function $2^{-N}\sum_\alpha e^{\beta X_\alpha}$ is coming from configurations $\alpha$ where $X_{\alpha_1}^1 \ll N\sqrt{\log 2}\,\sigma_1$ and $X_{\alpha_1,\alpha_2}^2 \ll N\sqrt{\log 2}\,\sigma_2$. In these regions, the energy levels are so dense that no individual configuration carries a macroscopic weight. This changes when $\beta$ crosses $\beta_1^{\mathrm{cr}}$, where the main contribution to the partition function is coming from energy levels $X^1_{\alpha_1}$ which "freeze" close to the maximal possible value, but not yet the $X^2_{\alpha_1,\alpha_2}$. The latter start to freeze at $N\sqrt{\log 2}\sigma_2$ only for $\beta \geq \beta^{cr}_2$. We make these considerations now more precise.

The Gibbs measure at inverse temperature $\beta > 0$ of $\alpha = (\alpha_1, \alpha_2)$, $1 \leq \alpha_i \leq 2^{N/2}$, is given by

$$P_{\omega,\beta,N}(\alpha) \stackrel{\text{def}}{=} 2^{-N} \frac{\exp[\beta X^1_{\alpha_1} + \beta X^2_{\alpha_1,\alpha_2}]}{Z_{\omega,\beta,N}} .$$

This is a random element (the randomness coming from $\omega$) in the set of probability measures on the configurations $\alpha = (\alpha_1, \alpha_2)$. We denote the set of these configurations again by $\Sigma_N$. As in the last lecture, we often regard this as a point process $(P_{\omega,\beta,N}(\alpha))_{\alpha \in \Sigma_N}$ on the interval $(0,1)$, where all the points sum up to 1. There are now three regimes for the temperature parameter: $\beta < \beta^{cr}_1$, $\beta^{cr}_1 < \beta < \beta^{cr}_2$, $\beta > \beta^{cr}_2$. As in the previous lecture, we will not discuss the critical cases $\beta = \beta^{cr}_1$ and $\beta = \beta^{cr}_2$. The high-temperature case $\beta < \beta^{cr}_1$ is easy: The outcome is exactly the same as in the high-temperature REM case. In particular, the individual contributions of single configurations $\alpha$ are asymptotically negligible:

$$(9.7) \qquad\qquad \sup_\alpha P_{\omega,\beta,N}(\alpha) \to 0, \text{ a.s.}$$

This is proved in the same way as Proposition 8.5.

(9.7) remains true also in the case $\beta^{cr}_1 < \beta < \beta^{cr}_2$, as will become clear below, but now, something interesting is happening with the first marginal distribution

$$P^1_{\omega,\beta,N}(\alpha_1) := \sum_{\alpha_2} P_{\omega,\beta,N}(\alpha_1, \alpha_2).$$

The fact that the first part $X^1_{\alpha_1}$ of the total energy freezes close to the maximal possible value leads to the result that this marginal stays macroscopic in the $N \to \infty$ limit for some $\alpha_1$:

**Theorem 9.2.** *Let $\beta > \beta^{cr}_1$. Then the point process*

$$(9.8) \qquad\qquad \left(P^1_{\cdot,\beta,N}(\alpha_1)\right)_{1 \leq \alpha_1 \leq 2^{N/2}}$$

*on $(0,1)$ converges weakly to $\mathcal{N}(PPP(t \to xt^{-x-1}))$,*

$$x = x(\beta) \stackrel{\text{def}}{=} \frac{\sqrt{\log 2}}{\beta\sigma_1}.$$

*Proof.* We write the marginal distribution as

$$P^1_{\omega,\beta,N}(\alpha_1) = 2^{-N} \frac{\exp[\beta X^1_{\alpha_1} + \beta U^{(N)}_{\alpha_1}]}{Z_{\omega,\beta,N}},$$

where

$$U^{(N)}_{\alpha_1} \stackrel{\text{def}}{=} \frac{1}{\beta} \log \sum_{\alpha_2=1}^{2^{N/2}} e^{\beta X^2_{\alpha_1,\alpha_2}}.$$

First observe that (9.3) under the side conditions (9.4) and (9.5) is maximized for $\beta > \beta_1^{cr}$ at

$$(s_1, s_2) = \left( \sqrt{\log 2} \sigma_1, \min(\beta \sigma_2^2, \sqrt{\log 2} \sigma_2) \right).$$

For abbreviation, we set

$$\varsigma = \varsigma(\beta) \stackrel{\text{def}}{=} \min \left( \beta \sigma_2^2, \sqrt{\log 2} \sigma_2 \right).$$

From the discussion of the free energy given above we can conclude, roughly speaking, that with $\mathbb{P}$-probability close to 1, $P_{\omega,\beta,N}$ is concentrated on those $(\alpha_1, \alpha_2)$ with $X_{\alpha_1}^1 \approx N\sqrt{\log 2}\sigma_1$, $X_{\alpha_1,\alpha_2}^2 \approx N\varsigma$, and therefore $P_{\omega,\beta,N}^1$ is concentrated on those $\alpha_1$ with

$$\frac{1}{N\beta} \log \sum_{\alpha_2} e^{\beta X_{\alpha_1,\alpha_2}^2} \approx \frac{1}{N\beta} \log \left( 2^{N/2} \exp \left[ -\frac{\varsigma^2 N}{2\sigma_2^2} \right] \exp[\beta N \varsigma] \right)$$

$$= \frac{1}{2\beta} \log 2 - \frac{\varsigma^2}{2\beta \sigma_2^2} + \varsigma \stackrel{\text{def}}{=} b(\beta).$$

To phrase it precisely: There are a sequence $\varepsilon_N \to 0$, and a sequence $\Gamma_N \subset \Omega$ of events satisfying

$$\lim_{N \to \infty} \mathbb{P}(\Gamma_N) = 1, \quad \lim_{N \to \infty} \sup_{\omega \in \Gamma_N} P_{\omega,\beta,N}^1 \left( \left\{ \alpha_1 : \left| \frac{U_{\alpha_1}(\omega)}{N} - b(\beta) \right| \leq \varepsilon_N \right\} \right) = 1.$$

We leave it as an exercise to derive this from the considerations in the proof of Proposition 9.1. We write $I_N$ for the interval $[Nb(\beta) - N\varepsilon_N, Nb(\beta) + N\varepsilon_N]$.

Using this, we see that the point process (9.8) has the same weak limit as the one where we leave out all points with $U_{\alpha_1} \notin I_N$

We will now show that there exists a sequence $(a_N)$ of real numbers, satisfying

(9.9) $$\lim_{N \to \infty} \frac{a_N}{N} = \sqrt{\log 2} \sigma_1 + b(\beta),$$

such that the point process

$$(X_{\alpha_1}^1 + U_{\alpha_1}^{(N)} - a_N)_{\alpha_1 : U_{\alpha_1} \in I_N}$$

converges weakly to a $PPP(t \to \frac{\sqrt{\log 2}}{\sigma_1} \exp[-\frac{\sqrt{\log 2}}{\sigma_1} t])$. From this it follows that

$$(\exp(\beta(X_{\alpha_1}^1 + U_{\alpha_1}^{(N)} - a_N)))_{\alpha_1 : U_{\alpha_1} \in I_N}$$

converges to a $PPP(t \to x(\beta) t^{-x(\beta)-1})$, and then the theorem follows, after a justification that one can interchange the normalizing operation $\mathcal{N}$ with the $N \to \infty$ limit, which is left to the reader.

To prove the claim let $\phi \in C_0^+(\mathbb{R})$, and write $\mu_N$ for the distribution of $U_{\alpha_1}^{(N)}$. Remark that $\mu_N(I_N) \to 1$ if $\varepsilon_N \to 0$ sufficiently slowly. For any sequence $(a_N)$ satisfying (9.9), we have

$$\mathbb{E} \exp\left[ -\sum_{\alpha_1} 1_{U_{\alpha_1} \in I_N} \phi(X_{\alpha_1}^1 + U_{\alpha_1} - a_N) \right]$$

$$= \left\{ 1 - \frac{1}{\sqrt{\pi N \sigma_1^2}} \int dt (1 - e^{-\phi(t)}) \int_{I_N} \mu_N(du) \exp\left[ -\frac{(t - u + a_N)^2}{2\sigma_1^2 N} \right] \right\}^{2^{N/2}}$$

$$= \left\{ 1 - \frac{1 + o(1)}{\sqrt{\pi N \sigma_1^2}} \int dt (1 - e^{-\phi(t)}) e^{-t\sqrt{\log 2}/\sigma_1} \int_{I_N} \mu_N(du) \exp\left[ -\frac{(a_N - u)^2}{2\sigma_1^2 N} \right] \right\}^{2^{N/2}}.$$

For $u \in I_N$, we have $a_N - u \approx N\sqrt{\log 2}\sigma_1$, and therefore $\exp[-\frac{(a_N - u)^2}{2\sigma_1^2 N}] \approx 2^{-N/2}$. We can adjust $(a_N)$ slightly, still satisfying (9.9), and such that

$$\frac{1}{\sqrt{\pi N \sigma_1^2}} \int_{I_N} \mu_N(du) \exp\left[ -\frac{(a_N - u)^2}{2\sigma_1^2 N} \right] = 2^{-N/2} \frac{\sqrt{\log 2}}{\sigma_1} (1 + o(1)).$$

With this choice, we get

$$\lim_{N \to \infty} \mathbb{E} \exp\left[ -\sum_{\alpha_1} 1_{U_{\alpha_1} \in I_N} \phi(X_{\alpha_1}^1 + U_{\alpha_1} - a_N) \right]$$

$$= \exp\left[ -\frac{\sqrt{\log 2}}{\sigma_1} \int dt (1 - e^{-\phi(t)}) e^{-t\sqrt{\log 2}/\sigma_1} \right],$$

which proves the claim. $\qquad \square$

There is a crucial difference in the behavior of the Gibbs distribution $(P_{\omega,\beta,N}(\alpha))_{\alpha \in \Sigma_N}$, depending on whether $\beta_1^{\mathrm{cr}} < \beta < \beta_2^{\mathrm{cr}}$ or $\beta > \beta_2^{\mathrm{cr}}$, despite the fact that there is no transition in the behavior of the marginal $(P_{\omega,\beta,N}^1(\alpha_1))_{1 \leq \alpha_1 \leq 2^{N/2}}$. The case

$$(9.10) \qquad\qquad \beta_1^{\mathrm{cr}} < \beta < \beta_2^{\mathrm{cr}}$$

reveals in the simplest possible way an effect which in physics literature is phrased in the notion of a "pure state". In Statistical Mechanics, a pure state is an extremal Gibbs measure after having taken the thermodynamic limit. For mean-field type models, there is no proper notion of a thermodynamic limit, and therefore the term "pure state" does not make much sense. However, for the GREM, it is easily explained what really is meant. From the discussion of the free energy, we see that the conditional distribution of the second component given the first $P_{\omega,\beta,N}(\alpha)/P_{\omega,\beta,N}^1(\alpha_1)$ is essentially concentrated on those $\alpha_2$ with $X_{\alpha_1,\alpha_2}^2 \approx N\beta\sigma_2^2$. However, in our regime (9.10), with $\mathbb{P}$-probability close to 1, there are in any interval $J$ close to $N\beta\sigma_2^2$, and of length of order 1, still exponentially many $\alpha_2$ with $X_{\alpha_1,\alpha_2}^2 \in J$. It therefore easily follows in a similar way as in Proposition 8.5 that no individual configuration can carry a $P_{\omega,\beta,N}$-measure of more than exponentially small (in $N$) weight. In order to catch a "macroscopic" weight, we therefore have to lump together exponentially many of the $\alpha$'s. According to the Theorem 9.2, we can achieve this by simply taking the collections $E_{\alpha_1} \stackrel{\text{def}}{=} \{(\alpha_1, \alpha_2) : 1 \leq \alpha_2 \leq 2^{N/2}\}$, and obtain for these disjoint sets the Gibbs weights $P(E_{\alpha_1}) = P^1(\alpha_1)$, which are macroscopic for some $\alpha_1$.

The "pure states" are then just the collections $E_{\alpha_1}$. The notion is still a bit imprecise. We could as well take somewhat smaller sets, restricting to those $\alpha_2$ with $X^2_{\alpha_1,\alpha_2} \approx \beta\sigma_2^2 N$. There are "countably many" of these pure states, in contrast to "uncountably many" individual configurations. This sentence, of course, does not make any sense as there are for any $N$ only finitely many configurations anyway. However, in order to get a macroscopic weight of the Gibbs distribution, one has to take exponentially many individual configurations, which is naturally interpreted as "uncountable" in the $N \to \infty$ limit. In contrast, for any $\varepsilon > 0$, there exists $K(\varepsilon)$, not depending on $N$, such that with $\mathbb{P}$-probability $\geq 1 - \varepsilon$, there are $K(\varepsilon)$ of the $E_{\alpha_1}$, such that their union has Gibbs weight $\geq 1 - \varepsilon$. There is therefore good reasons to say that there are "countably many pure states" in the $N \to \infty$ limit.

There is a notion of so-called metastates, (see [44]), which makes this discussion formally more precise, but we will not go into that.

For $\beta > \beta_2^{\text{cr}}$, the situation changes and individual configurations get now a macroscopic weight, and can therefore be considered as "pure states". The precise statement is given in the following result.

**Theorem 9.3.** *The point process* $(P_{\cdot,\beta,N}(\alpha))_{\alpha \in \Sigma_N}$ *converges weakly to* $\mathcal{N}(\text{PPP}(t \to x(\beta)t^{-x(\beta)-1}))$, *where*

$$x(\beta) \stackrel{\text{def}}{=} \frac{\sqrt{\log 2}}{\sigma_2 \beta}.$$

**Remark 9.4.** *The somewhat surprising fact is that only* $\sigma_2$ *enters into the parameter of the Poisson point process. This means that the variance of the energy levels* $X^1_{\alpha_1} + X^2_{\alpha_1,\alpha_2}$ *is not of relevance, but only the variance of* $X^2_{\alpha_1,\alpha_2}$.

**Sketch of the Proof of Theorem 9.3** From Proposition 8.6 of the last lecture, we know that there are sequences $(a_N^1)$ and $(a_N^2)$ with

$$\mathcal{L}((X^1_{\alpha_1} - a_N^1)_{1 \leq \alpha_1 \leq 2^{N/2}}) \stackrel{w}{\to} \text{PPP}(t \to a_1 e^{-a_1 t}),$$

and for any $\alpha_1$

$$\mathcal{L}((X^2_{\alpha_1,\alpha_2} - a_N^2)_{1 \leq \alpha_2 \leq 2^{N/2}}) \stackrel{w}{\to} \text{PPP}(t \to a_2 e^{-a_2 t}),$$

with $a_i \stackrel{\text{def}}{=} \sqrt{\log 2}/\sigma_i$. It is then easy to see that

$$\mathcal{L}((X^1_{\alpha_1} + X^2_{\alpha_1,\alpha_2} - a_N^1 - a_N^2)_{1 \leq \alpha_1,\alpha_2 \leq 2^{N/2}}) \stackrel{w}{\to} \mathcal{L}((\xi_i^1 + \xi_{i,j}^2)_{0 \leq i,j}),$$

where $(\xi_i^1)_{i \in \mathbb{N}}$ is a $\text{PPP}(t \to a_1 e^{-a_1 t})$, and for any $i$ the point processes $(\xi_{i,j}^2)_{j \in \mathbb{N}}$ are $\text{PPP}(t \to a_2 e^{-a_2 t})$ which are independent, and also independent of $\xi^1$.

We then define the point process

$$\eta_{i,j} \overset{\text{def}}{=} \exp\left[\beta\xi_i^1 + \beta\xi_{i,j}^2\right] = \eta_i^1\eta_{i,j}^2,$$

where $\eta_i^1 \overset{\text{def}}{=} \exp[\beta\xi_i^1]$ and $\eta_{i,j}^2 \overset{\text{def}}{=} \exp[\beta\xi_{i,j}^2]$. $(\eta_i^1)_{i\in\mathbf{N}}$ is a PPP$(t \to x_1 t^{-x_1-1})$ and for any $i$ $(\eta_{ij}^2)_{j\in\mathbf{N}}$ is a PPP$(t \to x_2 t^{-x_2-1})$, where $x_i \overset{\text{def}}{=} \sqrt{\log 2}/\sigma_i\beta$. Theorem 9.3 then follows from the following result:

**Proposition 9.5.** *Assume $\beta > \beta_2^{\text{cr}}$. Then*

a)

$$\sum_{i,j} \eta_{i,j} < \infty, \text{ a.s.}$$

b) $\mathcal{N}((\eta_{i,j}))$ *has the same law as* $\mathcal{N}(\text{PPP}(t \to x_2 t^{-x_2-1}))$.

*Proof.* a) follows from the fact that the intensities $t \to x_i t^{-x_i-1}$ are integrable at 0 as $x_1, x_2 < 1$. We leave the details to the reader.

b) We apply the Proposition 8.7b) of the last lecture. As $\sqrt{\log 2}/\sigma_2 < \beta$, we have

$$\sum_i e^{a_2\xi_i^1} < \infty, \text{ a.s.}$$

Therefore, conditionally on $(\xi_i^1)$, we see that $(\xi_i^1 + \xi_{i,j}^2)_{i,j}$ is a shift of a PPP$(t \to a_2 e^{-a_2 t})$. Therefore, conditioned on $(\xi_i^1)$, i.e. on $(\eta_i^1)$, $\mathcal{N}((\eta_{i,j}))$ has the same law as $\mathcal{N}(\text{PPP}(t \to x_2 t^{-x_2-1}))$, the latter not depending on the realization of $(\xi_i^1)$. From this, the statement of the proposition follows.                                □

Of course, a proof of Theorem 9.3 from the above proposition still needs some work as one has to show that the normalization operation $\mathcal{N}$ commutes with taking the $N \to \infty$ limit. We however skip the proof of this technical point.

We now describe the GREM with finitely many levels, which is a straightforward generalization of the GREM with 2 levels, we have just described. Again, we take $2^N$ centered Gaussian random variables $X_\alpha$ which have variance $N$. We represent them as

$$X_\alpha = X_{\alpha_1}^1 + X_{\alpha_1,\alpha_2}^2 + \ldots + X_{\alpha_1,\ldots,\alpha_k}^k,$$

where $\alpha = (\alpha_1,\ldots,\alpha_k)$, $1 \le \alpha_i \le 2^{N/k}$, and where the $X_{\alpha_1,\ldots,\alpha_i}^i$ are centered Gaussian random variables with variance $\sigma_i^2 N$, where we assume

$$\sigma_1^2 > \sigma_2^2 > \ldots > \sigma_k^2, \quad \sum_i \sigma_i^2 = 1.$$

We again write $\Sigma_N$ for the set of configurations $\alpha$. Completely similar to the situation with two levels, we now have $k$ critical points:

$$\beta_i^{\text{cr}} \overset{\text{def}}{=} \frac{\sqrt{2\log 2}}{\sqrt{k}\sigma_i}.$$

For $\beta \leq \beta_1^{\mathrm{cr}}$, the free energy $f(\beta)$ is the same as for the REM, and for $\beta \geq \beta_k^{\mathrm{cr}}$, the free energy is linear. In between, the free energy is a polynomial of second order in $\beta$, with a decreasing curvature as $\beta$ increases. The special form is not of any importance for us, and we leave it as an exercise to calculate it.

Far more interesting are the following facts: Define

$$x_i(\beta) \stackrel{\text{def}}{=} \frac{\sqrt{2 \log 2}}{\sqrt{k}\sigma_i \beta},$$

so that

$$x_1(\beta) < x_2(\beta) < \ldots < x_k(\beta).$$

If for some $M \leq k$ one has $\beta \in (\beta_M^{\mathrm{cr}}, \beta_{M+1}^{\mathrm{cr}})$, $(\beta_{k+1}^{\mathrm{cr}} \stackrel{\text{def}}{=} \infty$, $\beta_0^{\mathrm{cr}} \stackrel{\text{def}}{=} 0)$ then

$$x_M(\beta) < 1 < x_{M+1}(\beta).$$

This implies that the Gibbs distribution $(P_{\omega,\beta,N}(\alpha))_{\alpha \in \Sigma_N}$ is concentrated on those configuration $\alpha \in \Sigma_N$ where for $i \leq M$, $X^i_{\alpha_1,\ldots,\alpha_i}$ is close to their maximal possible value $\sigma_i \sqrt{2 \log 2}/\sqrt{k}$, but not for $i > M$. The precise picture for the appropriate marginal distributions can be derived in the same way as in the two-stage case and we just describe the outcome.

We rephrase the $N \to \infty$ limit in terms of Poisson point processes, which is the formulation of Ruelle [57]: There are sequences $(a_N^i)_{N \in \mathbf{N}}$, $i = 1, \ldots, k$ such that

$$\left(X^1_{\alpha_1} - a_N^1\right)_{1 \leq \alpha_1 \leq 2^{N/k}} \stackrel{w}{\to} \mathrm{PPP}\left(t \to a_1 \mathrm{e}^{-a_1 t}\right),$$

with $a_1 \stackrel{\text{def}}{=} \sqrt{2 \log 2}/\sqrt{k}\sigma_1$, and for $2 \leq i \leq k$, and all $\alpha_1, \ldots, \alpha_{i-1}$

$$\left(X^i_{\alpha_1,\ldots,\alpha_i} - a_N^i\right)_{1 \leq \alpha_i \leq 2^{N/k}} \stackrel{w}{\to} \mathrm{PPP}\left(t \to a_i \mathrm{e}^{-a_i t}\right),$$

with $a_i \stackrel{\text{def}}{=} \sqrt{2 \log 2}/\sqrt{k}\sigma_i$. It is then not difficult to show that for any $l \leq k$

$$\left(\exp\left[\beta\left(X^1_{\alpha_1} + X^2_{\alpha_1,\alpha_2} + \ldots + X^l_{\alpha_1,\ldots,\alpha_l} - (a_N^1 + \ldots + a_N^l)\right)\right]\right)_{1 \leq \alpha_1,\ldots,\alpha_l \leq 2^{N/k}}$$
$$\stackrel{w}{\to} \left(\eta^1_{i_1} \eta^2_{i_1,i_2} \cdot \ldots \cdot \eta^l_{i_1,\ldots,i_l}\right)_{i_1,\ldots,i_l \in \mathbf{N}},$$

where the $\eta$ are Poisson processes with the following properties: For any $j \leq k$ and any natural numbers $i_1, \ldots, i_{j-1}$, $(\eta^j_{i_1,\ldots,i_j})_{j_l \in \mathbf{N}}$ is a $\mathrm{PPP}(t \to x_j t^{-x_j-1})$. We retain the convention that the point processes are ordered downwards:

$$\eta^j_{i_1,\ldots,i_{j-1},0} > \eta^j_{i_1,\ldots,i_{j-1},1} > \eta^j_{i_1,\ldots,i_{j-1},2} \cdots$$

The $\eta^j$ are independent for different $j$, and $(\eta^l_{i_1,\ldots,i_l})_{j_l \in \mathbf{N}}$ are independent for different $i_1, \ldots, i_{l-1}$. Of course, the constants $a^1_N + \ldots + a^l_N$ don't play any role after normalization. As $x_M < 1$, one has

$$\sum_{i_1,\ldots,i_M} \eta^1_{i_1} \eta^2_{i_1,i_2} \cdot \ldots \cdot \eta^M_{i_1,\ldots,i_M} < \infty,$$

(but not for $M + 1$). We define

$$\eta_{i_1,\ldots,i_M} \stackrel{\text{def}}{=} \eta^1_{i_1} \eta^2_{i_1,i_2} \cdot \ldots \cdot \eta^M_{i_1,\ldots,i_M}.$$

In the same way as Proposition 9.5, it follows that $\mathcal{N}((\eta_{i_1,\ldots,i_M})_{i_1,\ldots,i_M \in \mathbf{N}})$ has the same law as $\mathcal{N}(\mathrm{PPP}\left(t \to x_M t^{-x_M - 1}\right))$.

**Proposition 9.6.** *Let the marginal distribution be defined by*

$$P^{(M)}_{\omega,\beta,N}(\alpha_1,\ldots,\alpha_M) \stackrel{\text{def}}{=} \sum_{\alpha_{M+1},\ldots,\alpha_k} P_{\omega,\beta,N}(\alpha).$$

*Then*

$$\left(P^{(M)}_{\omega,\beta,N}(\alpha_1,\ldots,\alpha_M)\right)_{1 \leq \alpha_1,\ldots,\alpha_M \leq 2^{N/k}} \stackrel{w}{\to} \mathcal{L}\big(\mathcal{N}((\eta_{i_1,\ldots,i_M})_{i_1,\ldots,i_M \in \mathbf{N}})\big)$$
$$= \mathcal{L}\big(\mathcal{N}(\mathrm{PPP}(t \to x_M t^{-x_M - 1}))\big).$$

## 9.2. The Clustering Mechanism Connected with the GREM

The last proposition states that the limiting Gibbs measure (or better its marginal on the "pure states") does not carry any information about the internal hierarchical structure. However, the hierarchical structure defines an additional random clustering structure on the points of the point process, which is of crucial importance for understanding the Parisi-picture of the SK-model. We discuss this clustering structure for the "$N = \infty$ point processes" directly, leaving aside questions about a discussion of the convergence of the finite $N$ clustering to this limit picture.

The points of our "limiting Gibbs measure" $\mathcal{N}((\eta_{i_1,\ldots,i_M})_{i_1,\ldots,i_M \in \mathbf{N}})$ can be ordered downwards

$$\zeta_0 > \zeta_1 > \ldots$$

To each $j \in \mathbf{N}$ we can attach uniquely an $\varphi(j) = (i_1, \ldots, i_M) \in \mathbf{N}^M$, such that

$$\zeta_j = \frac{\eta_{i_1,\ldots,i_M}}{\sum\limits_{i_1,\ldots,i_M} \eta_{i_1,\ldots,i_M}}.$$

If $\pi_l : \mathbf{N}^M \to \mathbf{N}^l$ is the projection on the first $l$ components, $l < M$, then $\pi_l \circ \varphi$ defines in the usual way a partition $\mathcal{Z}_l$ of $\mathbf{N}$: $j, j' \in \mathbf{N}$ belong to the same set of the partition if $\pi_l(\varphi(j)) = \pi_l(\varphi(j'))$. We also use the notation $j \sim_l j'$. These partitions are random elements in the set $E$ of all partitions on $\mathbf{N}$. Partitions are the same as equivalence relations, and any relation on $\mathbf{N}$ can be regarded as an element of $\{0,1\}^{\mathbf{N} \times \mathbf{N}}$. It is not difficult to see that $E$ is a closed subset of $\{0,1\}^{\mathbf{N} \times \mathbf{N}}$, the latter being equipped with the product topology, and therefore $E$ is a compact set, which we can equip with its Borel $\sigma$-field. Our random partitions are then measurable mappings taking values in this measurable set, and defined on the probability space on which the point processes are defined. Evidently, the partitions become coarser if $l$ decreases. For two partitions $\mathcal{A}$ and $\mathcal{B}$ we write $\mathcal{A} \succ \mathcal{B}$ if any set in $\mathcal{B}$ is a subset of a set in $\mathcal{A}$. With this notion we have

$$\mathcal{Z}_1 \succ \mathcal{Z}_2 \succ \ldots \succ \mathcal{Z}_{M-1}.$$

What the above clustering does is simply the following: Start with original $k$-stage GREM. Then we order the $M$-th level individual marginal probabilities downwards, and record the clustering stemming from the original hierarchical structure. In the $N \to \infty$ limit, this produces the above random partitionings.

An astonishing fact is the following result:

**Theorem 9.7.** $\mathcal{N}((\eta_{i_1,\ldots,i_M})_{i_1,\ldots,i_M \in \mathbf{N}})$ and $(\mathcal{Z}_1, \mathcal{Z}_2, \ldots, \mathcal{Z}_{M-1})$ are independent.

The proof of the above result can be found in [8]. We will not repeat it here, but will give an informal argument (for $M = 2$) which reveals, how the special form of the intensity of the point process comes into play. In the next lecture we will describe the law of the clustering explicitly.

Let therefore $M = 2$. It is slightly more convenient to work with the "energy" point processes. So we take $(\xi_i^1)_{i \in \mathbf{N}}$ as a PPP$(t \to a_1 e^{-a_1 t})$, and for any $i$ $(\xi_{i,j}^2)_{j \in \mathbf{N}}$ as PPP$(t \to a_2 e^{-a_2 t})$ which are independent, and also independent of $\xi^1$, where $a_1 < a_2$. We then take the joint point process $(\xi_i^1 + \xi_{i,j}^2)_{i,j \in \mathbf{N}}$. Ordering the points downwards, this defines a random partitioning on $\mathbf{N}$ as before. Remark that

$$\sum_i \exp[a_2 \xi_i^1] < \infty, \ a.s.$$

Take an arbitrary (nonrandom) sequence $\mathbf{x} = (x_i)$ satisfying $\sum_i \exp[a_2 x_i] < \infty$. We consider the partitioning defined by the point process $(x_i + \xi_{i,j}^2)_{i,j \in \mathbf{N}}$.

If $t \in \mathbb{R}$, then the probability that there is a point of this point process in an "infinitesimal" interval $[t, t+h]$ is

$$\sum_i \mathbb{P}\left(\exists j \text{ with } x_i + \xi_{i,j}^2 \in [t, t+h]\right) = h \sum_i a_2 \exp\left[-a_2\left(t - x_i\right)\right]$$

$$= h a_2 \exp\left[-a_2 t\right] \sum_i \exp\left[a_2 x_i\right].$$

On the other hand, the probability that the two infinitesimal intervals $[t, t+h]$ and $[s, s+h]$ are both occupied with clustered points is

$$\sum_i \mathbb{P}\left(\exists j \text{ with } x_i + \xi_{i,j}^2 \in [t, t+h] \text{ and } \exists j' \text{ with } x_i + \xi_{i,j'}^2 \in [s, s+h]\right)$$

$$= h^2 \sum_i a_2 \exp\left[-a_2\left(t - x_i\right)\right] a_2 \exp\left[-a_2\left(s - x_i\right)\right]$$

$$= h^2 a_2^2 \exp\left[-a_2\left(t + s\right)\right] \sum_i \exp\left[2a_2 x_i\right].$$

Therefore, the probability that both intervals are occupied and clustered, conditioned on the event that both intervals are occupied is simply

$$\frac{\sum_i \exp\left[2a_2 x_i\right]}{\left(\sum_i \exp\left[a_2 x_i\right]\right)^2}$$

which is completely independent of where the intervals are! Similar expressions are obtained for more complicated clustering events, all of which not depending where the infinitesimal intervals are, but depend on the sequence $\mathbf{x}$. In a similar way, one can compute probabilities of more complicated events. Given $m_1, \ldots, m_k \geq 1$, $N = \sum_{j=1}^k m_j$ and $N$ infinitesimal intervals, one computes the conditional probability given that all the intervals are occupied of the event that the points of the first $m_1$ are clustered, the points of the second $m_2$ intervals are clustered, but not with the $m_1$ group, etc. as

$$\sideset{}{^*}\sum_{i_1,\ldots,i_k} \exp\left(\sum_{j=1}^k m_j a_2 x_{i_j}\right) \Big/ \left(\sum_i \exp(a_2 x_i)\right)^N,$$

where $\sum^*$ denotes the summation over different indices, and these probabilities are independent of where the intervals are. It therefore follows (although this is not quite a formal proof) that the clustering defined by $(x_i + \xi_{ij}^2)_{i,j \in \mathbb{N}}$ on $\mathbb{N}$ is independent of the point process defined by this random object (one should of course keep in mind that $(x_i + \xi_{ij}^2)_{i,j \in \mathbb{N}}$ contains more information than the point process defined by it). If the $x_i$ are random, namely given by $(\xi_i^1)$ then the clustering probabilities are obtained by taking the $\mathbb{E}$-expectation in the end. For instance, the probability that the first, third and fourth point of $(\xi_i^1 + \xi_{ij}^2)_{i,j \in \mathbb{N}}$ are clustered, but not with the second, is

$$\mathbb{E}\left(\frac{\sum_{i \neq j} \exp(3a_2 \xi_i^1) \exp(a_2 \xi_j^1)}{\left(\sum_i \exp(a_2 \xi_i^1)\right)^4}\right).$$

We will come back to these clustering probabilities in the next lecture where we will give them a new interpretation as the transition probabilities of a Markov process.

Coming now back to the $k$-level GREM and the clustering $(\zeta_1, \zeta_2, \ldots, \zeta_{M-1})$ defined by it, we say that two points of the point process have *overlap* $j/k$ if they are the same set of the partition $\zeta_j$ but not of $\zeta_{j+1}$. In particular, "pure states" have overlap $M/k$. From the above theorem we see that these overlaps, which are random, are independent of the Gibbs weights, which are random, too, of course.

**Summarizing:** The hierarchical structure of the GREM introduces the notation of overlaps of "pure states". So, despite of the fact that the hierarchical structure does not enter into the law of the point process of the Gibbs weights, it does enter into their overlap structure.

The law of the overlaps is entirely determined by the sequence $0 < x_1(\beta) < x_2(\beta) < \cdots < x_M(\beta) \leq 1$. This dependence is usually encoded in a function $q : [0,1] \to [0,1]$ which is the crucial parameter of the Parisi-theory of mean-field type spin-glasses.

For the GREM, this $q$-function is simply defined by

$$
\begin{aligned}
q(x) &\stackrel{\text{def}}{=} M/k \quad &&\text{if} \quad x \geq x_M \\
q(x) &\stackrel{\text{def}}{=} j/k \quad &&\text{if} \quad x_j \leq x < x_{j+1}, \ 1 \leq j \leq M-1, \\
q(x) &\stackrel{\text{def}}{=} 0 \quad &&\text{if} \quad x < x_1.
\end{aligned}
$$

(9.11)

In the GREM discussed here, this is of course just a trivial step function, defined by the $x_j(\beta)$, whose $\beta$-dependence is very simple. For the SK-model, the Parisi-theory predicts a continuous $q$-function with a very delicate $\beta$-dependence. It is for the moment not clear how to define a clustering mechanism with a continuous $q$-function, but we will explain that in the next section.

# Lecture 10:

# Markovian Clustering, Reshuffling, and a Self-Consistency Equation

## 10.1. A Continuous Time Markovian Clustering Process

We define a Markov process in continuous time $(\Gamma_t)_{t \geq 0}$ taking values in the compact set $E$ of partitionings of $\mathbb{N}$. Transitions are only allowed to coarser partitionings, i.e. for $s \leq t$, we have $\Gamma_s \prec \Gamma_t$. The description is most simply given in terms of the traces on finite subsets $I \Subset \mathbb{N}$. We denote the finite set of partitionings of $I$ by $E^I$. Then the trace of $(\Gamma_t)_{t \geq 0}$ on $E^I$ is Markovian itself and is given by the following infinitesimal generator (i.e. the so called $Q$-matrix) $(a_{\Gamma,\Gamma'}^I)_{\Gamma,\Gamma' \in E^I}$: If $\Gamma'$ is obtained from $\Gamma \in E_I$ by clumping exactly $k$ classes, $k \geq 2$, into one class, and $\Gamma$ has $N$ classes, then

$$(10.1) \qquad a_{\Gamma,\Gamma'}^I \stackrel{\text{def}}{=} \frac{1}{(N-1)\binom{N-2}{k-2}}.$$

In all other cases $a_{\Gamma,\Gamma}^I \neq 0$ ($\Gamma' \neq \Gamma$). One should remark that infinitesimally, there is no clumping of different groups into more than one new set, but more than two sets in $\Gamma$ can be clumped into a new one of $\Gamma'$. As usual, one puts

$$(10.2) \qquad a_{\Gamma,\Gamma}^I \stackrel{\text{def}}{=} - \sum_{\Gamma':\Gamma' \neq \Gamma} a_{\Gamma,\Gamma'}^I,$$

and the semigroup $(R_t^I)$ on $E^I$ is defined by

$$(10.3) \qquad R_t^I = \exp(tA^I),$$

$A^I = (a_{\Gamma,\Gamma'}^I)$.

There is a very simple description, based on the observation that also the total number of classes left forms a Markov process. If there are $N$ sets of the partition at the beginning, then the process stays there for an exponential time with expectations $1/(N-1)$. At this jumping time, $X_1$ classes are clumped, where

$$P(X_1 = k) = \frac{N}{N-1} \frac{1}{k(k-1)}, \quad 2 \leq k \leq N.$$

Conditioned on $\{X_1 = k\}$ the choice of the $k$ classes to be clumped is uniform among the $\binom{N}{k}$ possibilities. After this, the partition has $N - X_1 + 1$ classes, and the clumping proceeds in the same way. It is clear that after finitely many of these clumpings, everything is clumped together into one class which is the absorbing state of the Markov process. These properties can easily be checked using the generator.

A curious fact is that the semigroup $(R_t^I)$, $I$ finite, can be calculated completely explicitly:

**Proposition 10.1.** *Assume that $\Gamma \in E^I$ has $N$ classes and $\Gamma'$ is obtained from $\Gamma$ by clumping $m_1, m_2, \ldots, m_k \geq 1$ classes of $\Gamma$, with $\Sigma_j m_j = N$. Then*

$$
(10.4) \qquad R_u^I(\Gamma, \Gamma') = \frac{(k-1)!}{(N-1)!} e^{-(k-1)u} \prod_{i=1}^{k} g_{m_i}(u),
$$

*where $g_1(u) \stackrel{def}{=} 1$, and for $m \geq 2$*

$$
(10.5) \qquad g_m(u) \stackrel{def}{=} (m - 1 - e^{-u})(m - 2 - e^{-u}) \times \cdots \times (1 - e^{-u}).
$$

The proof is not complicated but a bit tricky, and can be found in [8].

We now describe the clustering process on $\mathbb{N}$. The first observation is that the above clustering semigroups defined on finite subsets $I \Subset \mathbb{N}$ have a consistency property: Let $I \subset J \Subset \mathbb{N}$ and the natural projection

$$
\pi_{J,I} : E^J \to E^I,
$$

obtained by restricting a partitioning of $J$ to $I$. Let $\Gamma$ be any element of $E^I$ and $\tilde{\Gamma}$ be an element of $E^J$ with $\pi_{J,I}(\tilde{\Gamma}) = \Gamma$. Furthermore, let $\Gamma'$ be a partitioning of $I$ with $\Gamma' \succ \Gamma$. Then

$$
a_{\Gamma, \Gamma'}^I = \sum_{\tilde{\Gamma}' : \pi_{J,I}(\tilde{\Gamma}') = \Gamma'} a_{\tilde{\Gamma}, \tilde{\Gamma}'}^J
$$

by simply checking: If $\Gamma$ has $N$ classes, $\Gamma'$ is obtained from $\Gamma$ by clumping $k$ classes, and $\tilde{\Gamma}$ has $\tilde{N}$ classes, then for any $0 \leq l \leq \tilde{N} - N$, there are $\binom{\tilde{N}-N}{l}$ possibilities to choose a collection of $k + l$ classes in $\tilde{\Gamma}$ such that when clumped agree on $I$ with $\Gamma'$. Therefore, the above claimed equation is the same as

$$
\frac{1}{(N-1)\binom{N-2}{k-2}} = \sum_{l=0}^{\tilde{N}-N} \binom{\tilde{N}-N}{l} \frac{1}{(\tilde{N}-1)\binom{\tilde{N}-2}{k+l-2}}
$$

which is proved by checking. The above consistency equation implies the corresponding equation for the exponentials and this allows us to define a semigroup $(R_u)$ on $(E, \mathcal{E})$: If $\Gamma \in E$ then $R_u(\Gamma, \cdot)$ is the unique probability measure on $(E, \mathcal{E})$ ($\mathcal{E}$ the Borel-$\sigma$-Algebra, which is induced by $\pi_I : E \to E^I$, $I \Subset \mathbb{N}$) having the property that for any $I \Subset \mathbb{N}$, $\Gamma' \in E^I$

$$
(10.6) \qquad R_u(\Gamma, \pi_I^{-1}(\Gamma')) = R_u^I(\pi_I(\Gamma), \Gamma').
$$

It is readily checked that $(R_u)$ is a Feller semigroup on $E$, and in the usual way one associates to it a right-continuous strong Markov process, which we denote by $(\Gamma_t)_{t \geq 0}$, starting at the trivial partitioning $\Delta$ of $\mathbb{N}$.

It should be remarked that $\Gamma_t$ is nontrivial for all $t > 0$, with infinitely many classes, all of which have infinitely many elements. The reader is invited to check that.

We give now an alternative description of the above continuous time clustering process which will directly reveal the relation with the GREM defined in the last lecture.

Given a partition $\Gamma = \{C_1, C_2, \dots\}$ of $\mathbb{N}$, and $u > 0$, we describe the transition kernel $S_u(\Gamma, \cdot)$ in the following way: To each set $C_i$ we attach a random label $y_{C_i} \in \mathbb{N}$ whose distribution depends on $u$. Then we clump the $C_i$ with the same label. This defines $S_u(\Gamma, \cdot)$, but we still have to describe the distribution of the $y_{C_i}$. There are two stages of the construction. We first take a $\mathcal{N}(PPP(t \to xt^{-x-1}))$ with $x = e^{-u}$. Ordering as usual the points downward we arrive at a random probability distribution with decreasing weights $\bar{\eta}_0 > \bar{\eta}_1 > \dots$. Conditioned on $(\bar{\eta}_i)$, the $y_C$, $C \in \Gamma$ are i.i.d. with this distribution.

**Proposition 10.2.**
$$S_u = R_u, \quad u \geq 0.$$

*Proof.* We can define $(S_u^I)_{u \geq 0}$ for $I \Subset \mathbb{N}$ in a natural way using exactly the same construction, but now with only finite partitions. From the construction, it is then evident that the consistency relations (10.6) are satisfied. It therefore suffices to prove $S_u^I = R_u^I$, $u \geq 0$ for any finite $I \Subset \mathbb{N}$.

If $\Gamma \in E^I$ has $N$ classes and $\Gamma'$ is obtained by clumping $m_1, \dots, m_k \geq 1$ classes, $\sum_{i=1}^k m_i = N$, then

$$S_u^I(\Gamma, \Gamma') = \mathbb{E}\Big( \sum_{i_1,\dots,i_k}^* \bar{\eta}_{i_1}^{m_1} \dots \bar{\eta}_{i_k}^{m_k} \Big)$$

where $\sum^*$ means summation over indices which are pairwise distinct.

For the calculation of this expression, we use the following lemma which we state in a more general form than needed.

**Lemma 10.3.** *Let $f : (0, \infty) \to \mathbb{R}^+$ be continuous and satisfy*

$$(10.7) \qquad \int_0^\infty f(t)\, dt = \infty, \qquad \int_1^\infty f(t)\, dt < \infty, \qquad \int_0^\infty t f(t)\, dt < \infty.$$

*Let $(\eta_i)_{i \in \mathbb{N}}$ be a $PPP(f)$, which we can realize by an infinite sequence of random variables $\eta_0 > \eta_1 > \dots$ converging to $0$ and satisfy $\sum_i \eta_i < \infty$. If we define $\bar{\eta}_i \overset{def}{=} \eta_i / \sum_i \eta_j$, then for any $m_1, \dots, m_k \in \mathbb{N} \setminus \{0\}$:*
$$(10.8)$$

$$\mathbb{E}\Big( \sum_{i_1,\dots,i_k}^* \bar{\eta}_{i_1}^{m_1} \dots \bar{\eta}_{i_k}^{m_k} \Big) = \int_0^\infty \dots \int_0^\infty dt_1 \times \dots \times dt_k \prod_{j=1}^k t_j^{m_j} f(t_j) \mathbb{E}\Big( \frac{1}{(\sum_{j=1}^k t_j + \sum_l \eta_j)^N} \Big),$$

*where $N = \sum_{j=1}^k m_j$.*

We postpone the proof of this Lemma for the moment and proceed with the proof of Proposition 10.2, where we have $f(t) = xt^{-x-1}$, $x = e^{-u} < 1$.

We prove that if $\Gamma \in E^I$ and $\Gamma'$ is obtained from $\Gamma$ by clumping $m_1, \dots, m_k \geq 0$ classes, $\Sigma m_i = N$, where $N$ is the number of classes in $\Gamma$, then $S_n^I(\Gamma, \Gamma')$ is given by the expression (10.4). Now

$$S_n^I(\Gamma, \Gamma') = \mathbb{E}\Big( \sum_{i_1, i_2, \dots, i_k}^* \bar{\eta}_{i_1}^{m_1} \dots \bar{\eta}_{i_k}^{m_k} \Big).$$

Using Lemma 10.3 and $(m_1 - 1)$-times partial integration for $t_1$, $(m_2 - 1)$-times partial integration for $t_2$ etc., we get

$$(10.9) \qquad S_u^I(\Gamma, \Gamma') = \frac{(k-1)!}{(N-1)!} \prod_{j=1}^k g_{m_j}(u) \mathbb{E}\Big( \sideset{}{^*}\sum_{i_1,\dots,i_k} \overline{\eta}_{i_1} \dots \overline{\eta}_{i_k} \Big).$$

The claim follows by induction on $N$. $N = 1$ is trivial. Assume $N \geq 2$ and at least one $m_j \geq 2$. Then (10.9) reduces the problem to the calculation of $\mathbb{E}(\sideset{}{^*}\sum_{i_1,\dots,i_k} \overline{\eta}_{i_1} \dots \overline{\eta}_{i_k})$ on which we can apply the induction hypothesis because $k < N$, and we get for this term $\exp(-(k-1)u)$. It remains to consider the case $N \geq 2$, $k = N$, i.e. $m_1 = \dots = m_N = 1$. Then $\Gamma' = \Gamma$ and we get

$$S_u^I(\Gamma, \Gamma) = 1 - \sum_{\Gamma' \neq \Gamma} S_u^I(\Gamma, \Gamma').$$

For all summand on the right-hand side, we have already proved the claim, and therefore, it follows also for $S_u^I(\Gamma, \Gamma)$, because,

$$\sum_{\Gamma'} R_u^I(\Gamma, \Gamma') = 1.$$

*Proof of Lemma 10.3.* We prove the lemma in the case where $f$ has compact support in $(0, \infty)$. In that case, of course, $\int_0^\infty f(t)\, dt < \infty$, so that the assumptions (10.7) are not all satisfied. However, we just have to replace the left-hand side of (10.8) by

$$\mathbb{E}\Big( \sideset{}{^*}\sum_{i_1,\dots,i_k} \overline{\eta}_{i_1}^{m_1} \cdot \dots \cdot \overline{\eta}_{i_k}^{m_k} ; Z \geq k \Big),$$

where $Z$ is the number of points of the point process, which is Poisson with expectation $\int_0^\infty f(t)\, dt$. The case with infinitely many points in which we are interested in follows by limiting argument. We chop $[0, \infty)$ into small intervals $I_j = [jh, (j+1)h)$, $j \in \mathbb{N}$. $h > 0$ will go to 0 in the end. For $i = (i_1, \dots, i_k)$, all indices different, we consider the event $B$ that all points of the point process are in different ones of the intervals. Clearly

$$\mathbb{E}\Big( \sideset{}{^*}\sum_i \overline{\eta}_{i_1}^{m_2} \overline{\eta}_{i_2}^{m_2} \cdot \dots \cdot \overline{\eta}_{i_k}^{m_k} ; Z \geq k \Big) = \mathbb{E}\Big( \sideset{}{^*}\sum_i \overline{\eta}_{i_1}^{m_1} \cdot \dots \cdot \overline{\eta}_{i_k}^{m_k} ; B, Z \geq k \Big) + O(h)$$

where $O(h)$ refers to $h \downarrow 0$. The first summand on the right-hand side equals

$$\sideset{}{^*}\sum_{j_1,\dots,j_k \in \mathbb{N}} \mathbb{E}\Big( \sideset{}{^*}\sum_i \overline{\eta}_{i_1}^{m_1} \cdot \dots \cdot \overline{\eta}_{i_k}^{m_k} 1_{\{\eta_{i_1} \in I_{j_1}, \dots, \eta_{i_k} \in I_{j_k}\}} ; B \Big).$$

Fixing $j_1, \dots, j_k \in \mathbb{N}$, all different, then on the event $B$, we have

$$\sideset{}{^*}\sum_i \overline{\eta}_{i_1}^{m_1} \dots \overline{\eta}_{i_k}^{m_k} 1_{\{\eta_{i_1} \in I_{j_1}, \dots, \eta_{i_k} \in I_{j_k}\}} = \frac{(j_1 h)^{m_1} \dots (j_k h)^{m_k}}{(\sum_{s=1}^k j_s h + S)^N} A(j_1, \dots, j_k) + O(h)$$

where $S$ is the sum of the points of the point process outside $I_{j_1} \cup \dots \cup I_{j_k}$ and $A(j_1, \dots, j_k)$ is the event that each of the intervals $I_{j_s}$, $1 \leq s \leq k$, contains exactly one point. $A(j_1, \ldots, j_k)$ and $S$ are independent, and

$$\mathbb{P}(A(j_1, \ldots, j_k)) = \prod_{s=1}^{k} f(j_s h) h^k + O(h^{k+1}).$$

Furthermore, as $h \to 0$, the law of $S$ approaches the law of $\sum \eta_i$, uniformly in $j_1, \ldots, j_k$. Therefore

$$\mathbb{E}\Big( \sum_i{}^* \bar{\eta}_{i_1}^{m_1} \ldots \bar{\eta}_{i_k}^{m_k}; Z \ge k \Big)$$

$$= \int_0^\infty dt_1 \ldots \int_0^\infty dt_k f(t_1) \ldots f(t_k) t_1^{m_1} \ldots t_k^{m_k} \mathbb{E}\Big( \frac{1}{(\sum\limits_{j=1}^{k} t_j + \sum \eta_j)^N} \Big),$$

as claimed. This proves the Lemma.                                    □

The identification of the two semigroups $(R_u)_{u \ge 0}$ and $(S_u)_{u \ge 0}$ relates the former directly to the GREM of Lecture 9. This is immediate for the two-stage GREM, described in the point process setting by $(\eta_{i_1}^1)_{i_1 \in \mathbb{N}}$, $(\eta_{i_1 i_2}^2)_{i_2 \in \mathbb{N}}$ which are PPP$(t \to x_1 t^{-x_1-1})$ and PPP$(t \to x_2 t^{-x_2-1})$ respectively, $0 < x_1 < x_2 < 1$. In Lecture 9, we have described how, by ordering the points of $\eta_{i_1 i_2} = \eta_{i_1}^1 \eta_{i_1 i_2}^2$ downwards, this produces a random partition $\mathcal{Z}$ on $\mathbb{N}$. If $N \in \mathbb{N}$ and $\Gamma$ is a partitioning of $I_N \overset{\text{def}}{=} \{0, 1, \ldots, N\}$ with $k$ classes having $m_1, \ldots, m_k$ elements, then the probability that the trace $\mathcal{Z}$ on $I_N$ equals $\Gamma$ had been calculated as

$$\mathbb{E}\Big( \sum_{i_1, \ldots, i_k}{}^* \prod_{j=1}^{k} (\eta_{i_j}^1)^{m_j x_2} / \big( \sum_i (\eta_i^1)^{x_2} \big)^N \Big).$$

Now, $((\eta_i^1)^{x_2})$ is a PPP$(t \to \frac{x_1}{x_2} t^{-x_1/x_2-1})$. Therefore, the above probability is simply

$$S_u(\Delta, \pi_N^{-1}(\Gamma)), \qquad e^{-u} = \frac{x_1}{x_2}.$$

As $N$ is arbitrary, we conclude that the distribution of $\mathcal{Z}$ is $S_u(\Delta, \cdot)$. ($\Delta$ was the trivial partitioning into single points). These considerations can be extended to the GREM with $M$ kernels (Theorem 2.2 of [8]).

**Theorem 10.4.** *The law of the clustering process of Theorem 9.7 looked "backwards":*

$$(\mathcal{Z}_{M-1}, \mathcal{Z}_{M-2}, \ldots, \mathcal{Z}_1)$$

*is $(S_{u_1} \otimes S_{u_2} \otimes \cdots \otimes S_{u_{M-1}})(\Delta, \cdot)$, where $e^{-u_i} = x_i/x_{i+1}$. (For Markov kernels $K_1, \ldots, K_m$ on $(E, \mathcal{E})$, we write $K_1 \otimes \cdots \otimes K_m$ for the kernel from $(E, \mathcal{E})$ to $(E^m, \mathcal{E}^m)$ given by $K_1(x, dy_1) K_2(y_1, dy_2) \ldots$).*

We will not give a detailed proof of this which can be found in [8].

The above theorem gives a convenient way to define the overlap of two "pure states" of the GREM which had been introduced at the end of the last lecture. We consider a continuous time Markov process $(\Gamma_t)_{t \geq 0}$ with transition kernels $(R_u)_{u \geq 0} = (S_u)_{u \geq 0}$ and $\Gamma_0 = \Delta$. If $l, l' \in \mathbb{N}$, we define $\tau_{l,l'}$ to be the clustering time

$$\tau_{l,l'} \stackrel{\text{def}}{=} \inf\{ t : l, l' \text{ in the same class of } \Gamma_t \}.$$

Then the overlap is simply $q(X_{l,l'})$, where $X_{l,l'} = x_M \exp(-\tau_{l,l'})$, and $q$ is the step function defined in (9.11). This definition of an overlap is however possible for any (nice) function $q : [0, x_M] \to [0, q_M]$, $0 < x_M < 1$, $0 < q_M \leq 1$.

**Summarizing:** The basic ingredients are two independent processes: $(\eta_l)_{l \in \mathbb{N}}$ which is a $\text{PPP}(t \to x_M t^{-x_M - 1})$, $0 < x_M < 1$ ordered downwards $\eta_0 > \eta_1 > \eta_2 > \ldots$, and the clustering process $(\Gamma_t)_{t \geq 0}$ on the set $E$ of partitionings of $\mathbb{N}$, $\Gamma_0 = \Delta$.

The basic parameter of the theory is a function $q : [0, 1] \to [0, q_M]$, $0 < q_M \leq 1$, which is nondecreasing and satisfies $q(x) = q_M$ for $x \in [x_M, 1]$. The overlap of two "pure states", i.e. just $l, l' \in \mathbb{N}$ is then given by $q(X_{l,l'})$. For the $k$-level GREM, $q$ is a step function of (9.11), but we will focus now on the case where $q$ is a homeomorphism $[0, x_M] \to [0, q_M]$.

## 10.2. The Cavity Field and Reshuffling

We next introduce the so-called *cavity field* which is of crucial importance for the (non-rigorous) analysis of the SK-model by the "cavity method".

Parameterized by the $q$-function, this is a family of random variables $(y_l)_{l \in \mathbb{N}}$. The family is independent of $(\eta_l)_{l \in \mathbb{N}}$, and conditioned on the clustering process, i.e. the family $(X_{l,l'})$, the law of $(y_l)$ is centered Gaussian with covariances

$$E(y_l y_{l'}) = q(X_{l,l'}).$$

It is not difficult to check that such a family of random variables exists. The family $(y_l)_{l \in \mathbb{N}}$ is called the *cavity field*.

The motivation to consider this cavity field is coming from the SK-model. There one is investigating the effect of adding an $(N+1)$-th spin $\sigma_{N+1}$ to the $N$-system. The newcomer changes the Hamiltonian by a contribution

$$\frac{1}{\sqrt{N+1}} \sum_{i=1}^{N} \sigma_i J_{i,N+1} \sigma_{N+1}.$$

The cavity field is then the random field

$$\left( \frac{1}{\sqrt{N+1}} \sum_{i=1}^{N} \sigma_i J_{i,N+1} \right)_{\sigma = (\sigma_1, \ldots, \sigma_N)},$$

which is independent of the $N$-system Hamiltonian, and centered Gaussian. The covariances are given by

$$E\left( \frac{1}{\sqrt{N+1}} \sum_{i=1}^{N} \sigma_i J_{i,N+1} \frac{1}{\sqrt{N+1}} \sum_{i=1}^{N} \sigma'_i J_{i,N+1} \right) = \frac{1}{N+1} \sum_{i=1}^{N} \sigma_i \sigma'_i \approx \frac{1}{N} \sum_{i=1}^{N} \sigma_i \sigma'_i.$$

The basic idea is that the $\sigma = (\sigma_i)_{1 \leq i \leq N}$ can be assembled into "pure states" much like in the GREM, which have a macroscopic Gibbs weight. Furthermore, the distribution of $(\sigma_i)$ inside a "pure state" $l$ are supposed to be essentially independent, possibly with means $m_i^{(l)}$, depending on the pure state. Furthermore, the overlaps between configurations in different pure states should depend only on these states and not on the individual configurations. For the SK-model, there is however no mathematical proof of such a behavior. The newcomer changes the Gibbs weights of the pure states. Summing out the $\sigma_{N+1}$-variable, a pure state $l$ having the Gibbs weight $\overline{\eta}_l$ ($\sum_l \overline{\eta}_l = 1$) should have Gibbs weight in the $(N+1)$-system:

$$\overline{\eta}_l \exp(\log \cosh(\beta y_l)) / \sum_{l'} \overline{\eta}_{l'} \exp(\log \cosh(\beta y_{l'})).$$

There is also a change of the weight inside the $N$-system by having to replace $\frac{1}{\sqrt{N}}$ by $\frac{1}{\sqrt{N+1}}$, but this has no effect on the Gibbs weight of the pure states for $N \sim \infty$. We put $\psi(x) = \log \cosh(\beta x)$. The above considerations suggest that we should look at the trippel

$$((\overline{\eta}_l), (X_{l,l'}), (y_l)),$$

where $(\overline{\eta}_l) = \mathcal{N}((\eta_l))$, $(\eta_l)$ a PPP($t \to x_M t^{x_M - 1}$), and $(X_{l,l'}), (y_l)$ as defined above, and the reshuffling obtained by replacing $\eta_l$ by $\eta_l \exp(\psi(y_l))$. For technical reasons, we assume that $\psi$ is bounded, although this excludes the natural SK-choice $\psi(x) = \log \cosh(\beta x)$. We proceed now with a sketch of a rigorous analysis of this reshuffling operation. We order these points again downwards, which amounts to defining a random permutation $\sigma$ of $\mathbb{N}$ such that

$$\eta_{\sigma(l)} \exp(\psi(y_{\sigma(l)}))$$

is decreasing in $l$.

Put

$$\hat{\eta}_l \stackrel{\text{def}}{=} \eta_{\sigma(l)} \exp(\psi(y_{\sigma(l)})) / \sum_k \eta_k \exp(\psi(y_\varepsilon)).$$

Somewhat surprisingly this reshuffling has no effect on the Gibbs weights and the overlaps.

**Proposition 10.5.** $((\hat{\eta}_l), (X_{\sigma(l),\sigma(l')}))$ *has the same law as* $((\overline{\eta}_l), (X_{l,l'}))$.

On the other hand, it is fairly evident that the joint law including the cavity field changes after the reshuffling operation: The reshuffling certainly will favour the cavity variables where $\psi(y_l)$ is large, i.e. where $y_l$ is large if $\psi$ is monotone. We begin the discussion in the simple case where the cavity field is i.i.d., i.e. where

$$q(x) = \begin{cases} 0 & x < x_M \\ 1 & x \geq x_M. \end{cases}$$

The sequence $((\eta_l, y_l))_{l \in \mathbb{N}}$ then defines a Poisson point process on $(0, \infty) \times \mathbb{R}$ with intensity

$$(t, y) \to xt^{-x-1}p_1(y), \qquad x = x_M(= x_1)$$

where $p_s(y) = (2\pi s)^{-1/2} \exp(-y^2/2s)$. We transform the point process to the point process

$$(\eta_l e^{\psi(y_l)}, y_l).$$

**Lemma 10.6.** *This is a Poisson point process on $(0, \infty) \times \mathbb{R}$ with intensity*

$$(t, y) \to xt^{-x-1}e^{x\psi(y)}p_1(y).$$

*Proof.* Let $f \in C^+((0, \infty) \times \mathbb{R})$. Then

$$E \exp\left(-\sum_l f(\eta_l e^{\psi(y_l)}, y_l)\right) = \exp\left(-\int_{(0,\infty) \times \mathbb{R}} (1 - e^{-f(te^{\psi(y)}, y)})xt^{-x-1}p_1(y)\, dt\, dy\right)$$

$$= \exp\left(-\int_{(0,\infty) \times \mathbb{R}} (1 - e^{-f(t,y)})xt^{-x-1}e^{x\psi(y)}p_1(y)\, dt\, dy\right).$$

$\square$

We will now discuss the more interesting multi-level case. The case of a continuous $q$-function requires a limiting procedure where the number of the levels is going to $\infty$. For details of this, see Section 4 of [8]. We explain here the reshuffling for the two-level case. Therefore, we take

$$q(x) = \begin{cases} 1 & x \geq x_2 \\ q_1 & x_1 \leq x < x_2, \qquad 0 < q_1 < 1, \\ 0 & x < x_1. \end{cases}$$

We can realize the point process $(\eta_l)_{l \in \mathbb{N}}$ as $\eta_{i_1}^1 \eta_{i_1 i_2}^2$, where $(\eta_{i_1}^1)_{i_1 \in \mathbb{N}}$ is a $\mathrm{PPP}(t \to x_1 t^{-x_1-1})$ and for any $i_1$ $(\eta_{i_1 i_2}^2)_{i_2 \in \mathbb{N}}$ is a $\mathrm{PPP}(t \to x_2 t^{-x_2-1})$. The cavity field, we can realize as $y_{i_1}^1 + y_{i_1 i_2}^2$, where the $y^1$ are centered Gaussian with variance $q_1$, and the $y^2$ are centered Gaussian with variance $1 - q_1$. The clustering process is encoded in this representation, so we don't have to consider it separately for the moment. The reshuffling operation is done by considering the points

$$(10.10) \quad e^{\psi_0}\eta_{i_1}^1 \eta_{i_1 i_2}^2 \exp(\psi(y_{i_1}^1 + y_{i_1 i_2}^2)) = e^{\psi_0}\eta_{i_1}^1 e^{\psi_1(y_{i_1}^1)}\eta_{i_1 i_2}^2 e^{\psi(y_{i_1}^1 + y_{i_1 i_2}^2) - \psi_1(y_{i_1}^1)},$$

where

$$(10.11) \qquad e^{x_2\psi_1(y)} = \int e^{x_2\psi(y+z)}p_{1-q_1}(z)\, dz$$

$$(10.12) \qquad e^{x_1\psi_0} = \int e^{x_1\psi_1(z)}p_{q_1}(z)\, dz.$$

We now apply Lemma 10.6 twice:

$$(10.13) \qquad (\eta_{i_1}^1 e^{\psi_1(y_{i_1}^1) - \psi_0}, y_{i_1}^1)_{i_1 \in \mathbb{N}}$$

is a $PPP((t,y) \to x_1 t^{-x_1 - 1} e^{x_1 \psi_1(y) - x_1 \psi_0})$, and for $i_1$ and $y_{i_1}$ fixed,

$$(10.14) \qquad (\eta_{i_1 i_2}^2 e^{\psi(y_{i_1}^1 + y_{i_1 i_2}^2) - \psi_1(y_{i_1}^1)}, y_{i_1 i_2}^2)_{i_2 \in \mathbb{N}}$$

is a $PPP((t,y) \to x_2 t^{-x_2 - 1} e^{x_2 \psi(y_{i_1}^1 + y) - x_2 \psi_1(y_{i_1}^1)})$. It is important to remark that the first marginal of (10.14) is simply a $PPP(x_2 t^{-x_2 - 1})$, and is independent of $y_{i_1}^1$. Summarizing the situation, we see that the law of the variables

$$(\eta_{i_1}^1 e^{\psi_1(y_{i_1}^1) - \psi_0}, \eta_{i_1 i_2}^2 e^{\psi(y_{i_1}^1 + y_{i_1 i_2}^2) - \psi_1(y_{i_1}^1)}, y_{i_1}^1, y_{i_1 i_2}^2)$$

can be realized by $(\tilde{\eta}_{i_1}^1, \tilde{\eta}_{i_1 i_2}^2)$ having the same law as $(\eta_{i_1}^1, \eta_{i_1 i_2}^2)$ and independent of these, "reshuffled" cavity variables $(\tilde{y}_{i_1}, \tilde{y}_{i_1, i_2})$, where $\tilde{y}_{i_1}$ are i.i.d. random variables with density $e^{x_1(\psi_1(y) - \psi_0)} p_{q_1}(y)$, and conditioned on $(\tilde{y}_{i_1})_{i_1 \in \mathbb{N}}$ the variables $(\tilde{y}_{i_1, i_2})_{i_1, i_2 \in \mathbb{N}}$ are independent, $\tilde{y}_{i_1, i_2}$ with density $e^{x_2(\psi(y_{i_1}^1 + y) - x_2 \psi_1(y_{i_1}^1))} p_{1 - q_1}(y)$. We can rephrase this in terms of the clustering process: In the above situation, order

$$\eta_{i_1}^1 \eta_{i_1 i_2}^2 \exp(\psi(y_{i_1}^1 + y_{i_1 i_2}^2))$$

downwards: $\hat{\eta}_0 > \hat{\eta}_1 > \ldots$. If $l \in \mathbb{N}$ corresponds to $(i_1(l), i_2(l))$, write $\hat{y}_l^1 = y_{i_1}^1$, $\hat{y}_l^2 = y_{i_1 i_2}^2$. Furthermore, define the overlaps

$$\hat{q}_{l,l'} = \begin{cases} q_1 & \text{if } i_1(l) = i_1(l') \\ 0 & \text{otherwise.} \end{cases}$$

**Proposition 10.7.** *The distribution of $(\mathcal{N}((\hat{\eta}_l)), (\hat{q}_{l,l'}), (\hat{y}_l^1, \hat{y}_l^2))$ is given in the following way: $\mathcal{N}((\hat{y}_l))$ is the normalization of a $PPP(t \to x_2 t^{-x_2 - 1})$. Independently of this, we choose a random partition $\mathcal{Z}$ of $\mathbb{N}$, distributed according to $R_u(\Delta, \cdot)$, $e^{-u} = x_1/x_2$. Then*

$$\hat{q}_{l,l'} = \begin{cases} q_1 & \text{if } l \sim_z l' \\ 0 & \text{otherwise.} \end{cases}$$

*For any class $C \in \mathcal{Z}$ we choose independent $\hat{y}_C$, distributed according to*

$$\exp(x_1(\psi_1(y) - \psi_0)) p_{q_1}(y) \, dy,$$

*and we put $\hat{y}_l^1 = \hat{y}_C$ if $l \in C$. Conditioned on $\mathcal{Z}$ and $(\hat{y}_C)_{C \in \mathcal{Z}}$, we choose the $(\hat{y}_l^2)$ independent with distribution*

$$\exp(x_2(\psi(\hat{y}_C + y) - \psi_1(\hat{y}_C))) p_{1 - q_1}(y) \, dy,$$

*if $l \in C$.*

It should be evident, how this procedure extends to the finite level case.

We describe now, without proofs for which we refer to [8], the true reshuffling operation in the case of a continuous $q$-function. For technical reasons, we restrict ourselves to some technical conditions:

(10.15)   $q$ is a homeomorphism:   $[0, x_M] \to [0, q_M]$, $q(x) = q_M$ for $x \geq x_M$,

where   $0 < x_M < 1$, $0 < q_M \leq 1$.

(10.16)   $\psi : \mathbb{R} \to \mathbb{R}$   is bounded and has four continuous and

bounded derivatives.

As already is evident in the two-stage model, the cavity field has to be considered along the whole tree. In the continuous setting, we therefore consider the cavity field $(y_l)_{l \in \mathbb{N}}$ as a sequence of stochastic processes $y_l(t)$, $0 \leq t \leq x_M$. The basic setting is therefore a probability measure $\mathbf{Q}$ on $\Xi := M_p(\mathbb{R}^+) \times T \times C([0, x_M], \mathbb{R})^{\mathbb{N}}$, endowed with the natural product $\sigma$-algebra, where $T$ is the space of right continuous $E$-valued clustering functions. $\mathbf{Q}$ is given as

$$\mathbf{Q} = (\mu \otimes \gamma) \otimes \Sigma$$

where $\mu$ is the law of a $\mathcal{N}(\mathrm{PPP}(t \to x_M t^{-x_M - 1}))$, $\gamma$ is the law on $T$ of our clustering process $(\Gamma_t)_{t \geq 0}$ starting in the trivial clustering $\Delta$, and $\Sigma$ is the kernel from $T$ to $C([0, x_M], \mathbb{R})^{\mathbb{N}}$ which attaches to a realization $(\Gamma_t)_{t \geq 0}$ of the clustering process the law of centered Gaussian processes $(y_l(x))_{l \in \mathbb{N}, 0 \leq x \leq x_M}$ with covariances

$$\int y_l(x) y_{l'}(x') \Sigma((\Gamma_t)_{t \geq 0}, dy) = q(x \wedge x' \wedge X_{l,l'})$$

where $X_{l,l'} = x_M \exp(-\tau_{l,l'})$, $\tau_{l,l'}$ being the clustering time of $l, l'$ under $(\Gamma_t)$. It is not difficult to see that this kernel is well defined, and therefore also the measure $\mathbf{Q}$.

The reshuffling operation is easily described: We multiply the $\bar{\eta}_l$ by $\exp(\psi(y_l(x_M))$, order these points downwards, which defines a permutation of $\mathbb{N}$, normalize and apply the permutation to the clustering and the $y_l$ processes. This defines in an evident way a mapping $\Phi : \Xi \to \Xi$, which induces a change of measure of $\mathbf{Q} : \mathbf{Q}\Phi^{-1}$. This law is connected to the following PDE for $f \in C_b^{1,2}([0, q_M] \times \mathbb{R})$:

$$\frac{\partial f}{\partial q} + \frac{1}{2} \frac{\partial^2 f}{\partial y^2} + \frac{x(q)}{2} \left( \frac{\partial f}{\partial y} \right)^2 = 0$$

on $(0, q_M) \times \mathbb{R}$,

$$f(q_M, y) = \psi(y),$$

where $x(q)$ is the inverse mapping of $q(x)$ on $(0, q_M)$. One can prove that this has a unique solution. We set

$$m(q, y) = \frac{\partial f}{\partial y}(q, y).$$

For $N \in \mathbb{N}$, let $\mathcal{H}_N$ be the $\sigma$-field on $\Xi$ generated by $((\bar{\eta}_l), (\Gamma_t)_{t \geq 0}, (y_l)_{0 \leq l \leq N})$.

**Theorem 10.8.** *On* $\mathcal{H}_N$, *one has*

$$\mathbf{Q}\Phi^{-1} = J_N \cdot \mathbf{Q},$$

*where* $J_N$ *is defined in the following way. For* $0 < x \leq x_M$ *take the trace of* $T_{\log \frac{x_M}{x}}$ *on* $\{0, \ldots, N\}$ *and select a point in every class. Define* $L_N(x) \subset \{0, \ldots, N\}$ *to be this subset of representatives, which we do in such a way that the set changes only when the partitioning of* $\{0, \ldots, N\}$ *changes (which happens only a finite number of times). Then*

$$J_N = \exp\left\{ \int_0^{x_M} \sum_{l \in L_N(x)} xm(q(x), y^l(x)) \, dy^l(x) - \frac{1}{2} \int_0^{x_n} \sum_{l \in L_N(x)} x^2 m^2(q(x), y^l(x)) \, dq(x) \right\}.$$

The proof of this theorem is given in [8] (Th. 4.3).

### 10.3. A Consistency Equation for $q$

The reshuffling operation is basic to define an operation on the $q$-function. We can rigorously define this operation but the motivation is coming from non-rigorous considerations in the cavity approach for the SK-model, and we try to explain this first.

We consider a system with $N$ spins and assume that $\Sigma_N$ can be partitioned (approximately) into "pure states" $l$ which have Gibbs weights $\overline{\eta}_l$, ordered downwards, $(\overline{\eta}_l) = \mathcal{N}((\eta_l))$, where $(\eta_l)$ is a PPP($t \to x_M t^{-x_M - 1}$). Furthermore these pure states carry an overlap structure

$$q_{l,l'} = q(x_M e^{-X_{l,l'}}) \approx \frac{1}{N} \sum_{j=1}^{N} \sigma_j \sigma_j'$$

if $\sigma$ is in pure state $l$, and $\sigma'$ in $l'$, $l \neq l'$, where $q$ is the "true" $q$-function, to be determined. A newcomer $\sigma_{N+1}$ is now overtoppling the Gibbs weights. Remark that the Hamiltonian of the $(N + 1)$-system is

$$H_{N+1}(\sigma) = \frac{1}{\sqrt{N+1}} \sum_{1 \leq i < j \leq N+1} J_{ij} \sigma_i \sigma_j$$

$$= \sqrt{1 - \frac{1}{N+1}} \, H_N(\sigma) + \frac{1}{\sqrt{N+1}} \sigma_{N+1} \sum_{i=1}^{N} J_{i,N+1} \, \sigma_i.$$

The factor $\sqrt{1 - \frac{1}{N+1}}$ shouldn't cause any reshuffling in the $N \to \infty$ limit, but the second summand does. Remark that it is independent of $H_N$. It is plausible that we should get the reshuffling by summing out the $\sigma_{N+1}$-variable which amounts to multipling the $\eta_l$ by $\exp(\psi(y_l))$ and normalizing, where $y_l = \frac{1}{\sqrt{N}} \sum_{i=1}^{N} \sigma_i J_{i,N+1}$ and $\psi(x) = \log \cosh(\beta x)$. (We can of course replace $\frac{1}{\sqrt{N+1}}$ by $\frac{1}{\sqrt{N}}$). This leads then to reshuffled Gibbs weights $(\overline{\eta}_l^{(R)})$ and overlaps $(q_{l,l'}^{(R)})$, where we reorder the labels such that the $(\overline{\eta}_l^{(R)})$ are again ordered downwards. As remarked in the last lecture, $((\overline{\eta}_l^{(R)}), (q_{l,l'}^{(R)}))$ has the same law as before reshuffling.

The crucial idea is now that

$$q_{l,l'}^{(R)} \approx \frac{1}{N+1} \sum_{i=1}^{N+1} \sigma_i \sigma_i'$$

if $\sigma$ is in the pure state $l$ (of the $(N+1)$-system) and $\sigma'$ in $l'$. By "ergodicity" of the pure states, this should be $\frac{1}{N+1} \sum_{i=1}^{N+1} \langle \sigma_i \sigma_i' \rangle_{l,l'}$, where $\langle \cdot \rangle_{l,l'}$ refers to taking the Gibbs expectation of two independent realizations $\sigma$, $\sigma'$, one in $l$ and one in $l'$. If $x \leq x_M$, we would then get

$$\sum_{l,l'} 1_{q_{l,l'}^{(R)} \geq q(x)} q_{l,l'}^{(R)} \overline{\eta}_l^{(R)} \overline{\eta}_{l'}^{(R)} = \sum_{l,l'} 1_{q_{l,l'}^{(R)} \geq q(x)} \overline{\eta}_l^{(R)} \overline{\eta}_{l'}^{(R)} \Big\langle \frac{1}{N+1} \sum_{i=1}^{N+1} \sigma_i \sigma_i' \Big\rangle_{l,l'}.$$

We now take also the $\mathbb{E}$-expectation. By symmetry, we can replace

$$\Big\langle \frac{1}{N+1} \sum_{i=1}^{N+1} \sigma_i \sigma_i' \Big\rangle_{l,l'}$$

by $\langle \sigma_{N+1} \sigma'_{N+1} \rangle_{l,l'} \approx \mathrm{th}(\beta y_l^{(R)}) \, \mathrm{th}(\beta y_{l'}^{(R)})$. Therefore, we arrive in the $N \to \infty$ limit at

$$\mathbb{E}\Big( \sum_{l,l'} 1_{q_{l,l'}^{(R)} \geq q(x)} \overline{\eta}_l^{(R)} \overline{\eta}_{l'}^{(R)} \, \mathrm{th}(\beta y_l^{(R)}) \, \mathrm{th}(\beta y_{l'}^{(R)}) \Big)$$

$$= \mathbb{E}\Big( \sum_{l,l'} 1_{q_{l,l'}^{(R)} \geq q(x)} q_{l,l'}^{(R)} \overline{\eta}_l^{(R)} \overline{\eta}_{l'}^{(R)} \Big)$$

$$= \sum_{l \neq l'} \mathbb{E}(q_{l,l'}^{(R)} 1_{q_{l,l'}^{(R)} \geq q(x)} \overline{\eta}_l^{(R)} \overline{\eta}_{l'}^{(R)}) + q(x_M) \mathbb{E} \sum_l \overline{\eta}_l^{(R)2}$$

$$= \mathbb{E}(q_{0,1} 1_{q_{0,1} \geq q(x)}) \sum_{l \neq l'} \mathbb{E}(\overline{\eta}_l \overline{\eta}_{l'}) + q(x_M) \mathbb{E}\Big( \sum_l \overline{\eta}_l^2 \Big)$$

where we have used the independence of the overlaps and the Gibbs weights under $\mathbb{P}$. Now

$$q_{0,1} = q(x_M e^{-X_{0,1}}),$$

where $X_{0,1}$ is exponentially distributed. Therefore

$$\mathbb{E}(q_{0,1} 1_{q_{0,1} \geq q(x)}) = \frac{1}{x_M} \int_x^{x_M} q(s) \, ds.$$

On the other hand, $\mathbb{E}(\sum_l \overline{\eta}_l^2) = 1 - x_M$, and therefore, we get

$$\mathbb{E}\Big( \sum_{l,l'} 1_{q_{l,l'}^{(R)} \geq q(x)} \overline{\eta}_l^{(R)} \overline{\eta}_{l'}^{(r)} \, \mathrm{th}(\beta y_l^{(R)}) \, \mathrm{th}(\beta y_{l'}^{(R)}) \Big)$$

$$= \int_x^{x_M} q(s) \, ds + q(x_M)(1 - x_M) = \int_x^1 q(s) \, ds.$$

It is therefore natural to define for any (nice) function $q$ a "reshuffled" function $q^{(R)}$ by minus the derivative of the l.h.s. of the above equation. For technical reasons, we have assumed that $\psi$, which in the SK-context should be $\log\cosh(\beta x)$, is bounded, and we assume now also that it is symmetric. As

$$\mathrm{th}(\beta x) = \frac{(\log\cosh(\beta x))'}{\sup_x(\log\cosh(\beta x))'} ,$$

it is natural to replace $\mathrm{th}(\beta x)$ in the above expression by $\psi'(x)/\|\psi'\|_\infty$. Starting with a function $q$ satisfying the assumptions of the last subsection, we define

$$q^{(R)}(x) = \frac{-1}{\|\psi'\|_\infty^2}\frac{d}{dx_0}E^{\mathbf{Q}\Phi^{-1}}\left(\sum_{l,l'}\eta_l\eta_{l'}1_{X_{l,l'}\geq x_0}\psi'(y^l(x_M))\psi'(y^{l'}(x_M)))\right)\Bigg|_{x_0=x\wedge x_M}.$$

Of course, it has to be proved that this is well-defined. We can in fact represent the right-hand side with the help of the reshuffling objects defined in the last subsection. To this end, we define time inhomogeneous transition kernels $R_{x_0,x_1}$, $0\leq x_0\leq x_1\leq x_M$, on $\mathbb{R}$ as the transition kernels of the solution of the following stochastic differential equation

$$dy(x) = dM(x) + x\,m(q(x),y(x))\,dq(x),$$

$M(x) = B(q(x))$, $B$ being a standard Brownian motion.

The results of the last subsection imply the following result. For the proof we refer to [8]:

**Theorem 10.9.** *For $h$ bounded and measurable and $x_0\in[0,x_M]$*

$$E^{\mathbf{Q}\Phi^{-1}}\left(\sum_{l,l'}\eta_l\eta_{l'}1_{X_{l,l'}\geq x_0}h(y^l(x_M))h(y^{l'}(x_M)))\right)$$

$$= \int\limits_{x_0}^{x_M} R_{0,x}(R_{x,x_M}h)^2(0)\,dx + (1-x_M)R_{0,x_M}(h^2)(0).$$

*If we apply this to $h(x) = \frac{\psi'(x)}{\|\psi'(x)\|_\infty}$, we arrive at*

$$q^{(R)}(x) = \frac{1}{\|\psi'\|_\infty^2}R_{0,x\wedge x_M}(m(q(x\wedge x_M),\cdot)^2)(0),$$

$x\in[0,x_M]$.

**Proposition 10.10.** *$q^{(R)}$ is continuous increasing, with values in $[0,1]$, satisfies $q(0)=0$ and is constant on $[x_M,1]$, and is a homeomorphism $[0,x_M]\to[0,q_M^{(R)}]$, $q_M^{(R)} = q^{(R)}(x_M)$.*

In conclusion: We have defined in our abstract setup an operation on $q$-functions. In the SK-context, the fixed point equation

$$q = q^{(R)}$$

would be the consistency equation considered in the spin-glass literature.

We have however not addressed the most serious question, namely to prove that the SK-model can be described as $N \to \infty$ in some sense which has to be made precise by our abstract setup which essentially is developed here from the GREM.

On a more technical level, there are also a number of open problems:

- Relax the condition that $\psi$ is bounded. In particular include the case $\psi(x) = \log \cosh(\beta x)$.
- Prove that $q = q^{(R)}$ has a unique solution. In particular prove that there is also a unique $x_M < 1$.

# References

[1] M. Aizenman, J. Lebowitz, and D. Ruelle. Some rigorous results on the Sherrington-Kirkpatrick spinglass model. *Comm. Math. Phys.*, 116:527, 1988.

[2] N. Alon, J. Spencer, and P. Erdös. *The probabilistic method.* John Wiley & Sons, New York, 1992.

[3] P. Antal. *Trapping problem for the simple random walk.* Dissertation ETH, No 10759, 1994.

[4] P. Antal. Enlargement of obstacles for the simple random walk. *Ann. Probab.*, 23(3):1061–1101, 1995.

[5] G. Ben Arous and A.F. Ramirez. Asymptotic probabilities in the random saturation process. *Ann. Probab.*, 28(4):1470–1527, 2000.

[6] D. Boivin. Weak convergence for reversible random walks in random environment. *Ann. Probab.*, 21(3):1427–1440, 1993.

[7] E. Bolthausen. Localization of a two-dimensional random walk with an attractive path interaction. *Ann. Probab.*, 22:875–918, 1994.

[8] E. Bolthausen and A.-S. Sznitman. On Ruelle's probability cascades and an abstract cavity method. *Comm. Math. Phys.*, 197:247–276, 1998.

[9] J. Bricmont and A. Kupiainen. Random walks in asymmetric random environments. *Comm. Math. Phys.*, 142(2):345–420, 1991.

[10] A.A. Chernov. Replication of a multicomponent chain, by the "lightning mechanism". *Biophysics*, 12:336–341, 1962.

[11] F. Comets. A spherical bound for the Sherrington-Kirkpatrick model. Hommage à A. Meyer et J. Neveu. *Comm. Math. Phys.*, 236:103–108, 1996.

[12] F. Comets, N. Gantert, and O. Zeitouni. Quenched, annealed and functional large deviations for one-dimensional random walk in random environment. *Probab. Theory Relat. Fields*, 18:65–114, 2000.

[13] F. Comets and J. Neveu. The Sherrington-Kirkpatrick model of spin glasses and stochastic calculus: the high temperature case. *Comm. Math. Phys.*, 166:549–564, 1995.

[14] J.R.L. de Almeida and D.J. Thouless. Stability of the Sherrington-Kirkpatrick solution of a spin glass model. *J. Phys. A: Math. Gen.*, 11.

[15] A. Dembo, Y. Peres, and O. Zeitouni. Tail estimates for one-dimensional random walk in random environment. *Comm. Math. Phys.*, 181:667–683, 1996.

[16] B. Derrida. Random energy model: an exactly solvable model of disordered systems. *Phys. Rev.*, B 24:2613–2626, 1981.

[17] M. Donsker and S.R.S. Varadhan. On the number of distinct sites visited by a random walk. *Comm. Pure Appl. Math.*, 32:721–747, 1979.

[18] N. Dunford and J.T. Schwartz. *Linear operators*, volume I. Wiley, New York, 1988.

[19] R. Durrett. *Brownian motion and martingales in analysis.* Wadsworth, Belmont CA, 1984.

[20] R. Durrett. *Probability: Theory and Examples.* Wadsworth and Brooks/Cole, Pacific Grove, 1991.

[21] S.M. Ethier and T.G. Kurtz. *Markov processes.* John Wiley & Sons, New York, 1986.

[22] I. Fatt. The network model of porous media, III. *Trans. Amer. Inst. Mining Metallurgical, and Petroleum Engineers*, 207:164–177, 1956.

[23] J. Fröhlich and B. Zegarlinski. Some comments on the Sherrington-Kirkpatrick model of spin glasses. *Comm. Math. Phys.*, 112:553–566, 1987.

[24] E. Gardner and B. Derrida. The probability distribution of the partition function of the random energy model. *J. Phys.*, A 22:1975–1981, 1989.

[25] R.R. Hall. A quantitative isoperic inequality in $n$-dimensional space. *J. reine angew. Math.*, 428:161–176, 1992.

[26] F. den Hollander and G.H. Weiss. *Aspects of trapping in transport processes*. In: *Contemplary problems in Statistical Physics*, G.H. Weiss ed., SIAM, Philadelphia, 1994.

[27] B.D. Hughes. *Random walks and random environments*, volume 2. Clarendon Press, Oxford, 1996.

[28] V.V. Jikov, S.M. Kozlov, and O.A. Oleinik. *Homogenization of differential operators and integral functionals*. Springer, Berlin, 1994.

[29] S.A. Kalikow. Generalized random walk in a random environment. *Ann. Probab.*, 9:753–768, 1981.

[30] H. Kesten. A renewal theorem for random walk in a random environment. *Proc. Symposia Pure Math.*, 31:66–77, 1977.

[31] H. Kesten. The limit distribution of Sinai's random walk in random environment. *Physica A*, 138:299–309, 1986.

[32] H. Kesten, M.V. Kozlov, and F. Spitzer. A limit law for random walk in a random environment. *Compositio Mathematica*, 30(2):145–168, 1975.

[33] C. Kipnis and S.R.S. Varadhan. A central limit theorem for additive functionals of reversible Markov processes and applications to simple exclusions. *Comm. Math. Phys.*, 104:1–19, 1986.

[34] S. Kirkpatrick. Classical transport in random media: scaling and effective-medium theories. *Phys. Rev. Letters*, 27(25):1722–1725, 1971.

[35] W. Kirsch and F. Martinelli. Large deviations and the Lifschitz singularity of the density of states of random Hamiltonians. *Comm. Math. Phys.*, 89:27–40, 1983.

[36] S.M. Kozlov. The method of averaging and walks in inhomogeneous environments. *Russian Math. Surveys*, 40(2):73–145, 1985.

[37] U. Krengel. *Ergodic theorems*. Walter de Gruyter, Berlin, 1985.

[38] N.V. Krylov. An inequality in the theory of stochastic integrals. *Theor. Prob. Appl.*, 16(3):438–448, 1971.

[39] H.J. Kuo and N.S. Trudinger. Linear elliptic difference inequalities with random coefficients. *Mathematics of Computation*, 55(191):37–58, 1990.

[40] G.F. Lawler. Weak convergence of a random walk in a random environment. *Comm. Math. Phys.*, 87:81–87, 1982.

[41] M. Ledoux. *Isoperimetry and Gaussian analysis*. Ecole d'Eté de Probabilités de St. Flour 1994. Lecture Notes in Math., volume 1648, Springer, Berlin, 1987, 165–294.

[42] A. De Masi, P.A. Ferrari, S. Goldstein, and W.D. Wick. An invariance principle for reversible Markov processes. Applications to random motions in random environments. *J. Statist. Phys.*, 55(3-4):787–855, 1989.

[43] S.A. Molchanov. *Lectures on random media.* Ecole d'Eté de Probabilités de St. Flour XXII-1992, editor P. Bernard. Lecture Notes in Math., volume 1581, Springer, Berlin, 1994.

[44] C.M. Newman. *Topics in disordered systems.* Lectures in Mathematics ETH Zürich. Birkhäuser, Basel, 1997.

[45] E. Olivieri and P. Picco. On the existence of thermodynamics for the random energy model. *Comm. Math. Phys.*, 96:125–144, 1984.

[46] S. Olla. *Homogenization of diffusion processes in random fields.* Ecole Polytechnique, Palaiseau, 1994.

[47] G. Papanicolaou and S.R.S. Varadhan. *Boundary value problems with rapidly oscillating random coefficients.* in "Random Fields", J. Fritz, D. Szasz editors, Janyos Bolyai series, North-Holland, Amsterdam, 1981, 835–873.

[48] G. Papanicolaou and S.R.S. Varadhan. *Diffusion with random coefficients. Statistics and probability: essays in honor of C.R. Rao.* G. Kallianpur, P.R. Krishnajah, J.K. Gosh, eds., North Holland, Amsterdam, 1982, 547–552.

[49] A. Pisztora and T. Povel. Large deviation principle for random walk in a quenched random environment in the low speed regime. *Ann. Probab.*, 27(3):1389–1413, 1999.

[50] A. Pisztora, T. Povel, and O. Zeitouni. Precise large deviation estimates for one-dimensional random walk in random environment. *Probab. Theory Relat. Fields*, 113:191–219, 1999.

[51] T. Povel. Confinement of Brownian motion among Poissonian obstacles in $\mathbb{R}^d$, $d \geq 3$. *Probab. Theory Relat. Fields*, 114:177–205, 1999.

[52] M. Reed and B. Simon. *Methods of modern mathematical Physics*, volume I-IV. Academic Press, New York, 1972, 1975, 1978, 1979.

[53] S.I. Resnick. *Extreme Values, regular variation, and point processes.* Springer, New York, 1987.

[54] P. Revesz. *Random walk in random and non-random environments.* World Scientific, Singapore, 1990.

[55] H.B. Rosenstock. Random walks with spartaneous emission. *J. Soc. Indus. Appl. Math.*, 9:169–188, 1961.

[56] H.B. Rosenstock. Luminescent emission from an organic solid with traps. *Phys. Rev.*, 187:1166–1168, 1969.

[57] D. Ruelle. A mathematical reformulation of Derrida's REM and GREM. *Comm. Math. Phys.*, 108:225–239, 1987.

[58] L. Saloff-Coste. *Lectures on finite Markov chains*, volume 1665. Ecole d'Eté de Probabilités de Saint Flour, P. Bernard, ed., Lectures Notes in Mathematics, Springer, Berlin, 1997.

[59] U. Schmock. Convergence of the normalized one dimensional Wiener sausage path measure to a mixture of Brownian taboo processes. *Stochastics*, 29:171–183, 1990.

[60] Ya.G. Sinai. The limiting behavior of a one-dimensional random walk in a random environment. *Theory Prob. Appl.*, 27(2):247–258, 1982.

[61] F. Solomon. Random walk in a random environment. *Ann. Probab.*, 3:1–31, 1975.

[62] A.S. Sznitman. On the confinement property of Brownian motion among Poissonian obstacles. *Comm. Pure Appl. Math.*, 44:1137–1170, 1991.

[63] A.S. Sznitman. Brownian asymptotics in a Poissonian environment. *Probab. Theory Relat. Fields*, 95:155–174, 1993.

[64] A.S. Sznitman. *Brownian motion, obstacles and random media*. Springer, Berlin, 1998.

[65] A.S. Sznitman. Slowdown and neutral pockets for a random walk in random environment. *Probab. Theory Relat. Fields*, 115:287–323, 1999.

[66] A.S. Sznitman. Slowdown estimates and central limit theorem for random walks in random environment. *J. Eur. Math. Soc.*, 2:93–143, 2000.

[67] A.S. Sznitman. On a class of transient random walks in random environment. *Ann. Probab.*, 29(2):723–764, 2001.

[68] A.S. Sznitman and M.P.W. Zerner. A law of large numbers for random walks in random environment. *Ann. Probab.*, 27(4):1851–1869, 1999.

[69] M. Talagrand. The Sherrington-Kirkpatrick model: a challenge for mathematicians. *Probab. Theory Related Fields*, 110:109–176, 1998.

[70] D.E. Temkin. One-dimensional random walks in a two-component chain. *Soviet Math. Dokl.*, 13(5):1172–1176, 1972.

[71] D.J. Thouless, P.W. Anderson, and R.G. Palmer. Solution of 'solvable model of a spin glass'. *Philosophical Magazine*, 35:593–601, 1977.

[72] M. von Smoluchowski. Versuch einer mathematischen Theorie der koagulations-kinetischen Lösungen. *Z. Wahrsch. Phys. Chemie*, 92:129–168, 1918.

[73] M.P.W. Zerner. Lyapunov exponents and quenched large deviation for multidimensional random walk in random environment. *Ann. Probab.*, 26:1446–1476, 1998.

[74] M.P.W. Zerner and F. Merkl. A zero-one law for planar random walks in random environment. To appear in *Ann. Probab.*, 2001.

# Index